Power of UncertAInty

Anand Venkatanarayanan is a cybersecurity researcher with over three decades of experience advising corporates and at-risk individuals on emerging threats. His multidisciplinary work spans financial modelling, cybercrime and money-laundering networks, information warfare, and geopolitics.

Anand is widely recognized for his writing and public commentary on cybersecurity, large-scale frauds, information warfare and digital rights, and for bridging deep technical insight with regulatory, financial and societal perspectives. He is also the author of *The Art of Conjuring Alternate Realities*, a widely read book on information warfare.

You can connect with him on:

LinkedIn: https://www.linkedin.com/in/anandvenkatanarayanan/
X: @iam_anandv

Srinath Sridharan is a strategic counsel with over three decades of experience advising leading corporates, boards and promoters. His work spans leadership and governance transformation, succession planning, business scaling and institutional resilience. He is also a CEO coach.

He is the author of *Family and Dhanda*, a widely acclaimed book on Indian family businesses, and *Life's 101 Greatest Secrets*. His earlier co-authored works include *Time for Bharat*, *Newsprint to Heartprint* and *Reimagining ESG*.

Srinath is a Visiting Fellow at the Observer Research Foundation, and writes and speaks extensively on governance, leadership and the intersection of technology, institutions and society.

You can connect with him on:

LinkedIn: https://in.linkedin.com/in/srinathsridharan
X: @ssmumbai
Instagram: @AuthorSrinath

Power of UncertAInty

The Battle for Privacy, Identity and Sovereignty

Anand V. and Srinath Sridharan

RUPA

Published by
Rupa Publications India Pvt. Ltd 2026
161-B/4, Gulmohar House,
Yusuf Sarai Community Centre,
New Delhi 110049

Sales centres
Bengaluru Chennai
Hyderabad Kolkata Mumbai

P-ISBN: 978-93-7646-858-4
E-ISBN: 978-93-7646-572-9

First impression 2026

10 9 8 7 6 5 4 3 2 1

Printed in India

For all of us—
whose lives are shaped by systems we did not design,
and whose freedom must not depend on them.

Contents

Authors' Note

The world order is no longer stable; it has become chaotic and disruptive. *Pax Americana** appears to be receding, underscored by the United States' inability to subdue Iran in the third Gulf War. Much of the volatility we see in the world today is amplified by digital technologies and deeply interlinked industrial supply chains. These have become the lifeblood of modern economies, the foundation of decision-making and essential tools for innovation. For India, a nation with ambitions to lead in the digital century, managing this transition represents an unparalleled opportunity—and an equally daunting challenge.

This book was born out of our shared conviction that India has a role to play in redefining how nations manage digital technologies and supply chains in the 21st century. The phrase 'data is the new oil' captures its economic significance but overlooks a crucial truth: data is also a toxic asset, akin to spent nuclear fuel.

Emerging technologies such as artificial intelligence, quantum computing and large language models (LLMs) are actively shaping societal behaviour, economic models and even sovereign identities. For India, the stakes are especially high. Not yet part of the middle-income group of countries, but driven by far greater aspirations, can it earn a seat at the table by innovating at scale—or will it

***Pax Americana* refers to the period of relative global stability, security and economic integration that followed the Second World War, underpinned by the political, military, economic and cultural dominance of the United States. Drawing from the Latin *pax* (peace), the term echoes earlier formulations such as *Pax Romana*, and denotes not the absence of conflict, but an international order shaped by a prevailing power that sets the rules of trade, security alliances, financial systems and global institutions.

simply resign itself to being a user and net importer, surrendering those aspirations?

The title, *Power of UncertAInty*, reflects the evolving nature of power in this new order. Intelligence is no longer confined to human institutions; it is increasingly embedded in systems that operate at scale, across borders and often beyond effective oversight. These systems are opaque, adaptive and capable of shaping economic outcomes, political behaviour and social stability. In this sense, they represent both a blessing and a curse: a blessing in their ability to enhance decision-making, drive innovation and manage complexity; a curse in their opacity, their potential to concentrate power and the risks they introduce when control is diminished. As uncertainty becomes a defining feature of the age, power increasingly resides in the ability to navigate and shape it.

For India, this is not a question of digital progress alone, but of strategic positioning in a contested technological landscape. The choices made now will determine whether it builds sovereign control over intelligence, data and infrastructure, or remains structurally dependent on external ecosystems. While the country has made significant strides in digital public infrastructure and access, that is only the foundation. The harder task ahead is institutional: aligning regulation, supply chains and state capacity with the realities of technological power, and ensuring that sovereignty, privacy and identity are not incrementally ceded.

It is a reminder that the time to act is now—and that, with the right strategies, India can create a unique blueprint for the next 75 years: one that safeguards its sovereignty while positioning it as a beacon of innovation on the global stage.

In our discussions with policymakers, regulators and industry leaders, one question consistently emerged: *how can India develop institutional and technological capabilities that govern responsibly while fostering innovation without smothering the freedom of citizens?*

This book aims to answer that question. It is a call to reimagine governance not as a system of controls, but as an ecosystem of capabilities, powered by supply chains that preserve the balance between citizens and the state.

We believe the world stands at the cusp of what we call a new subprime moment—not of mortgages or credit derivatives, but of intelligence itself.

It is the greatest FOMO of our times: a race to deploy, to own and to monetize artificial intelligence before fully understanding its risks, ethics or consequences. Just as subprime mortgages were sold on the illusion of endless growth, today's AI gold rush is fuelled by the illusion of endless intelligence—the belief that more data, more models and more speed will automatically yield more progress. Beneath the glitter, however, lies leverage of a different kind: dependence on invisible systems we neither built nor control.

At the heart of this book lies the belief that data and technology are not merely resources but responsibilities. Their power resides not in accumulation or dominance, but in thoughtful, ethical application. For India, this represents a small window of opportunity to redefine its governance systems, strengthen its regulatory frameworks and set a global benchmark for ethical innovation.

We began this work from a shared unease: a sense that while India's digital progress has been impressive, its frameworks for trust, equity and accountability are still catching up. The technologies we celebrate are reshaping not only markets and governance, but imagination itself. They are altering how we see ourselves and what we expect of the state, the market and one another. These changes demand not just regulatory clarity but moral clarity.

Our effort here is to bring together the technical, the institutional and the foundational. We have tried to look beyond the daily noise of policy headlines and towards the deeper architecture of sovereignty—how it is built, how it frays and how it might be renewed.

The book speaks of data and AI, but ultimately it is about power: who holds it, who delegates it and who must be trusted to use it well.

It is also important to recognize that while many of the issues related to artificial intelligence, digital infrastructure, regulatory frameworks and geopolitical responses are evolving at extraordinary speed, the broader questions and challenges examined in these pages will continue to remain highly relevant.

We know this is unfinished work. The questions we raise will outlive us, as they should. Future generations will debate and refine them in ways we cannot imagine. Our hope is simply that this book contributes, even modestly, to that ongoing conversation—and that it inspires those who govern, regulate, build and question—to do so with both confidence and conscience.

Anand and **Srinath**

Introduction

Two Assassinations and the Dependency on Pax Americana

An intelligence operative we once met described how stunned he was to see Israeli tech in action, as far back as 2011.* He said, 'All I had to give them is [sic] a phone number and they were able to accurately track my movements over the next 30 days within 100 metres accuracy in the heart of Delhi. Never once after that demonstration, I ever carried my phone to meet people to have sensitive conversations.' The operative was not formally trained in digital counter-surveillance techniques, but had an instinctive understanding of them, having served the country for over three decades in hostile environments.

The basis of this tracking technique lies in the architecture of telecom networks. To support international roaming, protocols evolved to exchange location and routing information (i.e., which network serves a given SIM). Because the design of these networks is largely based on trust, any operator who can peer with a subscriber's network can send signalling messages that reveal the approximate cell-tower location to which a device is currently attached.

The operative understood what former US National Security Agency and Central Intelligence Agency Director Michael Hayden spoke about when discussing metadata at a university forum: 'We kill people based on metadata. But that is not what we do with

*For a detailed account of how location tracking operates, see 'Surveillance Secrets', *Lighthouse Reports*, 14 October 2025, https://www.lighthousereports.com/investigation/surveillance-secrets/. Accessed on 22 April 2026.

metadata.'* Metadata, or data about data, is often as valuable as the data itself, and in some cases, reveals far more than data.

For instance, consider a (somewhat) hypothetical case of a prankster sending a bomb threat to an airline from a newly created email account. A typical law-enforcement investigation would first seek the last IP address from which the email was sent, then approach the telecom service provider to identify the customer to whom that IP address was assigned, and then finally attempt to locate the individual using the address provided when the account was opened.

If that approach hits a roadblock, investigators can request the device identifier, examine other devices logged in from the same IP address and identify accounts potentially linked via common phone numbers. Only when multiple correlations emerge would they seek court authorization to access the email account, citing these linkages and requesting a deeper investigation. More often than not, courts honour such requests, as do the service providers offering the email service.

Digital trails are thus left everywhere by every one of us, irrespective of what we do, and they are nearly impossible to mask unless one is specially trained in deception skills. Such skills are not taught to ordinary citizens, yet there is a public expectation that intelligence services—which routinely use digital trails to unmask conspirators and conduct their own analyses—would themselves be trained in counter-surveillance and deception.

There are two specific reasons why these expectations exist:

- If one were to understand the construct of a nation-state, it may be useful to think of it as a higher-order structure than a family, village, city or tribe. Because it is a higher-order

*'Video Clip of Former Director of NSA and CIA: "We Kill People Based on Metadata"', *Just Security*, 12 May 2014, https://www.justsecurity.org/10318/video-clip-director-nsa-cia-we-kill-people-based-metadata/. Accessed on 25 May 2026.

structure composed of lower-order aggregates, it is inherently more fragile than those aggregates. Fragility is the price paid for benefits such as unity of purpose and the capacity to act collectively as a society. Law-enforcement agencies, intelligence services and the armed forces exist to manage this fragility through overt and covert means.

- Conducting covert actions, both within the state's boundaries and beyond, in furtherance of national interests and preventing other states from doing the same, is the core purpose of intelligence services. Given 15 years of evidence showing how digital trails can be exploited in myriad ways, one would reasonably assume that digital capabilities exist on both the offensive and defensive fronts.

The recent incident in which a serving Research and Analysis Wing officer was indicted by a US court* for allegedly planning an assassination on US soil suggests an astonishingly poor standard of digital security practice. Publicly available court records reveal that:

1. The operative had a personal Gmail account prior to induction into the covert service.
2. He stored his entire life's data on Google Drive, including his personal information, identity and address proofs, salary slips, tax returns and even his offer letter from the covert service.
3. This Gmail account was linked to an Indian phone number.
4. He created a second Gmail account to plan his activities in the US and linked it to another Indian phone number.
5. He used this second phone number to create a WhatsApp account but accessed it from the same device linked with his personal Gmail account.
6. In some instances, he mixed up the two Gmail accounts,

*'Vikash Yadav', *FBI Most Wanted*, https://www.fbi.gov/wanted/counterintelligence/vikash-yadav. Accessed on 22 April 2026.

sending emails from his second account instead of the personal Gmail account.

In effect, the operative possessed no essential digital tradecraft and lacked even a basic understanding of how to compartmentalize his life. Even a rookie would begin their digital counter-intelligence by:*

a. Purchasing a new laptop.
b. Creating a new virtual machine that wipes itself on shutdown and leaves no traces on the disk.
c. Using off-the-record (OTR) messaging tools.
d. Hiding within the ambient noise of the internet.

The operative did not do any of this and appears not even to have been aware of such practices. The irony is that his CV, which was stored on Google Drive, claimed expertise in all aspects of cybersecurity.

Internal commentary within India on this incident has been muted, and some have suggested that digital sovereignty offers one solution: if an Indian equivalent of an email and storage service had existed, the provider would not have honoured—in this case, as Google did—the court order granting the FBI full access to both of the operatives' Google accounts.

It is a bizarre suggestion, because sovereignty, from the point of view of a nation-state, means the capability to act with the tools at its disposal and to manage the consequences of whatever happens next. The operative chose Google's services, not the reverse. One can safely infer from the court documents that his training did not even include modules on counter-intelligence techniques of the kind routinely practised by ordinary drug traffickers and teenage ransomware gangs in the West. Put simply, the covert service did not have a digital DNA and consequently suffered the humiliation

*grugq, 'Ignorance Is Strength', *Hacker OPSEC*, June 2013, https://grugq.github.io/blog/2013/06/13/ignorance-is-strength/. Accessed on 22 April 2026.

of having that deficiency exposed in a public document.

Even if one sets aside the morality and legality of targeted assassinations (all nation-states do it), the lack of a digital DNA within the intelligence services points to a disturbing question: when covert services—endowed with far greater freedom of action and funding than rule-bound law-enforcement agencies, and whose only mission is to safeguard the interests of the Indian state and manage its fragility—lack such basic capacities, what use are loud pronouncements of digital sovereignty and self-sufficiency on the civilian side?

In the cyber domain, offensive operations and defensive technologies are two sides of the same coin. Progress in one advances the other; so too does degradation.

If one compares the absence of digital tradecraft in this offensive operation with the complete lack of digital-security discipline practised by the Iranian military during the short 12-day war with Israel, another disturbing lesson emerges: metadata can indeed be used to track, kill and potentially decapitate not only armed forces but also civilian leadership.

Iranian officials may have avoided using mobile phones to evade Israeli tracking, but signals intelligence is not confined to the phones of the targets themselves; it extends to those within their circles as well. Consider a recent experience of the authors when they met a high-ranking state law enforcement officer. The officer had a state-assigned chauffeur not merely to drive him around, but also an office attendant whom he would summon to carry his bag from his office to his car. He was always surrounded by attendants, helpers, bodyguards and personal secretaries.

Iranian officials were no different in that respect. Even before the war began, Israel was able to map the entire web of relationships linking these officials to their support staff. Unlike the officials, however, the support staff were not trained in digital security. They carried their digital devices wherever they went, including when they had to drive their superiors to top-secret locations, whose geographic coordinates could then be inferred through cluster-analysis algorithms.*

Some members of the support staff even had social-media accounts, where they regularly posted about their lives and communicated with extended family via WhatsApp. Given that WhatsApp's end-to-end encryption applies only to content and not to metadata, one can readily imagine a scenario in which the US sought specific data on these individuals under its national-security laws through a PEN register request or a Title III request in real time,† with which Meta would have had to comply but could never publicly disclose.

Data localization and similar mandates will not help here because the control plane of the infrastructure resides in the US and is not localized. Furthermore, none of the service providers, Meta included, has publicly explained how it would manage conflicts of a kinetic nature when the US becomes directly involved. The US did, after all, bomb Iran. Their denials, therefore, that WhatsApp is secure by virtue of end-to-end (E2E) encryption can be dismissed as self-serving and are, in that sense, not entirely credible.

For Israel to track the leadership of the Iranian armed forces and civilian establishment and conduct a red wedding—a mass decapitation campaign—all it would really require is continuous real-time location data from the support staff's mobile phones, whether through known vulnerabilities in the telecom signalling systems deployed by Iran, enriched by ancillary data from Meta (for example IP addresses, which refine location information, the

---

*'Targeting Iran's Leaders, Israel Found a Weak Link: Their Bodyguards', *The New York Times*, 30 August 2025, https://www.nytimes.com/2025/08/30/us/politics/israel-iran-assassination.html. Accessed on 22 April 2026.

†Meta, 'Government Requests for Data in the United States of America', *Meta Transparency Center*, https://transparency.meta.com/reports/government-data-requests/country/US/. Accessed on 25 May 2026.

groups they belong to, whom they message, how often, and the circles of relationships around them). This can then be further used to compromise such staff through coercion, blackmail or bribery.

Digital trails can lead to unexpected pathways that significantly influence the trajectory of a regime, and this extends beyond leadership and support staff to their extended families as well.

This is what General Hayden referred to when he observed that metadata is used for much more than killing people, even though it is used for that too.

Digital sovereignty, when considered from this perspective, simply means the following:

1. A nation-state cannot rely solely on critical digital infrastructure built and maintained by providers who could weaponize it against the state in times of conflict.
2. Control must not be merely a legal fiction (for example, a local office, data-localization norm, or local data centre), but also include observable and verifiable technical controls that are auditable.
3. Further, the architecture of the infrastructure must be explainable, auditable and customizable to local needs, rather than opaque or rigidly baked in.
4. The requirements in points 2 and 3 must extend across the entirety of the supply chain.
5. There must exist an industrial ecosystem capable of creating competing players who can deliver products and services on these lines, supported by a fiscal and monetary base adequate to sustain that ecosystem.
6. Finally, the state must possess the policy, governance and societal capacity to plan and execute these goals over the long term, without being distracted by short-term considerations.

Only when all these conditions are met can a nation-state achieve true digital sovereignty. But how does one create the societal base that would allow such an ecosystem to emerge?

There is only one sure-fire way: freedom for the citizenry to build and an environment in which public policymaking becomes thoughtful, evidence-based and meritocratic. Sovereignty, therefore, is not an end in itself but an emergent phenomenon of freedom, and should be understood as a cultural artefact.

Consider the origin story of WhatsApp. Jan Koum is Ukrainian, and his early memories of living in a totalitarian communist state—where everyone informed on everyone else and everything was eavesdropped upon and recorded—were so deeply traumatic that he vowed to create a product designed to prevent precisely that. That is why even its earliest design did not rely on central storage for messages, but instead used a 'store-from-the-sender, forward-to-the-receivers and then delete' architecture once all receivers had obtained the message.

It was a radical design even then, when most enterprise messaging applications stored data on central servers. By instinctively understanding that a mobile-era social network resides not in random strangers on the internet sending friend requests, but in people storing one another's phone numbers in contact lists—and by building a product that was private and intimate—Jan Koum and Brian Acton were eventually able to sell it to Facebook for US$19 billion, with a do-not-interfere (no ads) clause.

They later added end-to-end encryption, such that even the service provider could not see message content, by incorporating the Signal protocol. Some portion of the proceeds from the Facebook acquisition was then invested in the Signal Foundation to build and scale the Signal messenger, which, unlike WhatsApp, does not even retain metadata and is widely regarded as the global gold standard for private messaging.

Signal messenger thus offers one alternative model for building sovereign infrastructure. Consider the following facts:

1. The protocol itself is open, and anyone can adopt and use it in their own products.
2. There already exists sufficient knowledge, know-how, code, white papers, libraries and client and server code as open source to build an equivalent.
3. A fully auditable and verifiable protocol that prevents service providers from weaponizing it, coupled with an open codebase that allows anyone to verify the latest code running on their mobile devices, provides visibility across the supply chain.
4. The non-profit model means that no Big Tech company or government can dictate the technological or architectural choices in ways that compromise the core promise of end-to-end encryption with zero-metadata surveillance.

But Signal messenger is not what the now-dead Iranian generals used to conduct their affairs. They used WhatsApp, and that points to a more disturbing truth: the nation-states in which they lived did not create an environment that permitted individuals the freedom to be themselves; instead, they systematically repressed such freedom.

This does not, however, mean that state officials are indifferent to freedom for themselves. President Putin of the Russian Federation, for instance, famously carries his own toilet wherever he goes so that no other nation-state can analyse his excretions and infer details about his health and diet. Contrast that with the cyber-espionage attack on AIIMS in New Delhi, disguised as ransomware,* in which patient records—including the health data

*Milenkoski, Aleksandar, and Julian-Ferdinand Vögele, 'ChamelGang & Friends: Cyberespionage Groups Attacking Critical Infrastructure with Ransomware', *SentinelLabs*, 26 June 2024, https://www.sentinelone.com/labs/chamelgang-attacking-critical-infrastructure-with-ransomware/. Accessed on 22 April 2026.

of the prime minister and his Cabinet—were reportedly stolen. Yet, apart from suspending two junior NIC employees,* no systematic review of the implications appears to have been undertaken. One may therefore be forced to conclude, painfully, that talk of digital sovereignty in India is not yet taken seriously enough as a strategic tool or as a core sovereign capability.

The Indian state may build products like Aadhaar, but it will not address systematic defects in their design because doing so would puncture the comfort of believing it has arrived at its desired state of digital maturity. It is far more comfortable to speak of building a home-grown WhatsApp killer than to adopt Signal messenger across the entire country and government—including the intelligence services—by running our own servers for less than US$25 million a year. Taiwan's armed forces do precisely this.

Sovereignty is often mistaken for the crude emotional instinct of 'not invented or built by us', rather than understood as a much more nuanced set of capabilities that can either be built or acquired. This confusion stems from a structural problem: the Indian state does not believe that its citizenry can do things on its own and constantly attempts to second-guess their intent. Instead, it believes it must guide them, shape their path and often coerce them into a preferred approach, much like a parent telling a three-year-old to brush their teeth regularly.

This tendency often leads to the rise of elites and insiders, who then persuade the bureaucratic apparatus of the state that their way is the only way, and in doing so place the full weight of the state behind themselves. The result is the creation of national champions shielded not only from foreign competition but from domestic competition as well. The net result is the emergence of inferior

---

*'AIIMS Delhi Suspends Two System Analysts over Failing to Tackle Ransomware Attack', *The Economic Times*, 30 November 2022, https://economictimes.indiatimes.com/news/india/aiims-delhi-says-e-hospital-data-being-restored-services-still-on-manual/articleshow/95862820.cms. Accessed on 22 April 2026.

products and services, tailored narrowly to the local market and incapable of being exported elsewhere.

One side effect of this approach is the loss of geopolitical and global mercantile power and influence. Only when a nation-state can embed itself in the digital supply chains of other countries does it possess leverage when unforeseen events occur in the geopolitical arena.

These accumulated weaknesses are then weaponized in times of geopolitical conflict. Consider, for instance, the suspension of the Microsoft account of Nayara Energy after it was placed under US sanctions without notice. The absence of alternatives to Big Tech means that only legal measures to override such a suspension are available to the Indian state, because competitive substitutes do not exist. Fragility, therefore, becomes systemic and can only be managed through firefighting, even as the underlying vulnerabilities continue to harden.

How Current Geopolitical Alignments Restrict India's Sovereignty

Throughout the 21st century, numerous global commentaries and expert forecasts have projected that India could match or even surpass China as a major economic power. This view gained traction particularly in the early 2010s, when then Chief Economic Adviser Kaushik Basu projected that India might overtake China's growth rate within three to four years, citing robust domestic conditions and supportive IMF forecasts.* The optimism was echoed in a 2015 analysis by *IndiaSpend*, which noted that, on a purchasing power parity (PPP) basis, India could surpass China by 2016, while in real

*Press Trust of India (PTI), 'India May Surpass China's Growth Rate in 3–4 Years: Basu', *Mint*, 3 March 2011, https://www.livemint.com/Politics/ywJcc5E5dGfKBTqwnqgQTJ/India-may-surpass-China8217s-growth-rate-in-34-years-Ba.html. Accessed on 22 April 2026.

GDP terms, such a crossover could occur by 2032.*

The projection of India's economic rise was further reinforced by the broader BRICS framework coined by Goldman Sachs, which included India as a future global growth leader alongside Brazil, Russia and China.† The idea of an 'Indian Century' began to take root in strategic discourse, promoting the view that India could dominate global affairs in the latter half of the century because of its economic strength. In subsequent years, this optimism continued. At the 2019 World Economic Forum in Davos, former Reserve Bank of India Governor Raghuram Rajan stated that India was well positioned to eventually surpass China economically and take on a larger role in South Asia's development landscape.‡

India's development model was described by Larry Summers as the 'Mumbai Consensus', which he defined as 'the idea of a democratic developmental state, driven not by a mercantilist emphasis on exports, but a people-centred emphasis on growing levels of consumption and a widening middle class.' Parallel to its economic strides, India's strategic alignment with the United States has strengthened markedly since the early 2000s. During the Cold War, India belonged to the Non-Aligned Movement and leaned towards the Soviet bloc, but that continuity shifted after liberalization in the 1990s. A turning point arrived after 2004, when India and the US forged a strategic partnership featuring civilian nuclear collaboration, defence pacts and joint military exercises.

---

*Salve, Prachi, 'Can India's GDP Surpass China's? By 2032, Perhaps', *IndiaSpend*, 22 January 2015, https://www.indiaspend.com/can-indias-gdp-surpass-chinas-by-2032-perhaps-77817. Accessed on 22 April 2026.

†'Ask the Expert: BRICs and Investor Strategy', *Financial Times*, 6 November 2006, https://web.archive.org/web/20150425020625/http://www.ft.com/cms/s/2/c8192f64-6a66-11db-83d9-0000779e2340.html. Accessed on 22 April 2026.

‡'India to Surpass China, Be in Better Position to Help South Asia: Rajan', *Business Standard*, 22 January 2019, https://www.business-standard.com/article/economy-policy/india-to-surpass-china-be-in-better-position-to-help-south-asia-rajan-119012201094_1.html. Accessed on 22 April 2026.

India–US trade also improved steadily over time and by 2024–25,* the US was India's largest trading partner, with bilateral trade of US$131 billion and a surplus of US$41 billion in India's favour. Relations with the US also improved steadily, as India was seen as a counterweight to China's assertiveness. That dynamic changed dramatically after Donald Trump was elected President of the United States again.

Trump's worldview on trade, for nearly 50 years, had always bordered on the bizarre. Even as a real-estate developer, he viewed vendors on whom he depended to deliver goods and services as somehow ripping him off merely because his business made them rich. He extended the same zero-sum logic to international trade and came to see any nation-state with a trade surplus with the United States as self-serving.† Tariffs then became, in his view, an equalizing mechanism for correcting those surpluses. The reason this view is so bizarre is that it ignores the extraordinary privilege the US dollar enjoys within the global trading system.

As the pre-eminent global reserve currency, the US dollar does not have to fall in value against the currencies of countries with which the US runs a trade deficit. Countries sell their goods and services to the US and accept in return a piece of paper that the US Treasury can produce at near-zero cost because of the country's unmatched ability to attract capital flows. So, while it is true that many countries grew rich by trading with the US, the US grew much richer.[1]

However, the gains were not evenly distributed, as most of them accrued to the financial elite who were best positioned to

*'US Top trading Partner of India for 4th Straight Year', *Times of India*, April 16 2025, https://timesofindia.indiatimes.com/business/india-business/us-top-trading-partner-of-india-for-4th-straight-year-in-fy25-trade-gap-with-china-widens-to-99-2-bn/articleshow/120333409.cms

†Viser, Matt, 'Trump's Long History Hating Global Trade — and Loving Tariffs', *The Washington Post*, 9 April 2025, https://www.washingtonpost.com/politics/2025/04/09/trump-tariffs-history-stock-market/

capture these inflows. The widespread resentment this produced was one reason Trump was able to tap into, create the MAGA (Make America Great Again) movement, and return to power on a promise to fix the problem. Tariffs were not the right solution, but that did not concern him. In dealing with Trump, India made several foreign-policy mistakes:[2]

1. Trump has been obsessed with winning a Nobel Peace Prize, largely because Obama got one, and the Indian prime minister did not play to this obsession by ignoring Trump's claims of mediating the ceasefire after the three-day air war between India and Pakistan that followed Pahalgam.*
2. Foreign policy is not personal chemistry. Neither spectacle-driven events such as *Namaste Trump* and *Howdy Modi*, nor one-to-one meetings, are sufficient to yield long-term results.
3. India did not engage with Trump when he was out of power, thereby signalling that power alone is the currency that matters.
4. It embraced the US-driven sanctions regime beyond India's stated position of accepting only UN-mandated sanctions, thereby exposing itself to the current pressures.

These choices have produced the current dilemma, in which the US has not only imposed tariffs on India that are closer to sanctions,[3] but has also effectively put India's export trade to the US in play as a weapon through which to direct how India should use its earned dollars, where it should spend them, and how far it should open access to farm products—something no

---

*Neena Gopal, in her article 'End of the Kashmiri Summer: Is Pahalgam the Payback for the Jaffar Express Hijacking?', *The New Indian Express*, 23 April 2025, speculates whether the Pahalgam attack was a response to the Pakistani Army's perception that India was behind the Jaffar Express hijacking. See https://www.newindianexpress.com/web-only/2025/Apr/23/end-of-the-kashmiri-summer-is-pahalgam-the-payback-for-the-jaffar-express-hijacking for more details. Accessed on 25 May 2026.

prime minister could easily accept. Trump has also indicated that he would tax remittances to India, which would hurt not only lending but the capital account as well, by worsening the export–import imbalance.

The deterioration is especially harmful for India because its relationship with the US had developed in the context of strategic rebalancing against the coercive diplomacy and military pressure exerted by China on its borders, pressure aimed at constraining India's rise. As Manoj Kewalramani notes,* the conflict between India and China can be managed but not avoided because of three interlocking dynamics:

- First, both powers have overlapping spheres of influence.
- Second, China has a much larger economy and therefore far greater military power.
- Third, India is the swing power that will influence the outcome of the rivalry in an increasingly multipolar world.

That being the case, both the US and China see India as a key player with which to counter the other, but each seeks to do so by reducing India's autonomy to determine its own course of action, thereby constraining its sovereignty. India, meanwhile, is attempting to author its own rise as a regional power, but by playing both sides, it risks turning both powers against it.

Consider, for instance, China's restrictions on the export of fertilizers to India,[4] and its efforts to ensure that India does not move further up the electric-vehicle supply chain by withholding rare minerals and magnets.† It has also restricted[5] the movement

---

*Sengupta, Ramananda, 'India-China Ties: Don't Misread Thaw As Transformation', *StratNews Global*, 28 July 28 2025, https://stratnewsglobal.com/china/india-china-ties-dont-misread-thaw-as-transformation/. Accessed on 25 May 2026.

†Sachdev, Alisha, Sudhi Ranjan Sen, and Shruti Srivastava, 'China Rare Earth Curbs Choke India Automakers, Add to Tensions', *Bloomberg*, 6 June 2025, https://www.bloomberg.com/news/articles/2025-06-06/china-rare-earth-curbs-choke-production-by-indian-automakers-add-to-tensions. Accessed on 25 May 2026.

of engineers from its factories into India to train Indian companies in manufacturing iPhones, as part of both the government's *Make in India* policy and Apple's longer-term China+1 strategy to reduce dependence on China.

To summarize just how bizarre the current situation is:

- India saw itself as a peer competitor to China, but China did not. This perception served US interests, and therefore, relations between India and the US moved on an upward trajectory.
- Recognizing the potential threat, however, China opted for coercion through diplomatic, military and economic means, which pushed India closer to the US and made it more dependent on export surpluses with the US to offset its own trade deficit with China.
- Trump, however, viewed this as a strategic vulnerability to exploit through a sanctions-style regime perfected over time by the dollar's exorbitant privilege.
- This is now pushing India back towards China through efforts at normalization of ties that had deteriorated after the Galwan incident, in which 20 Indian soldiers were killed in hand-to-hand combat with the People's Liberation Army,* after which India lost access[6] to a significant portion of the territory it claims through the creation of buffer zones.

Compressed in the Middle (of everything and everyone)

Larry Summers' Mumbai Consensus envisaged a widening middle class, a growing economy, and a strong public sector to ensure prosperity not just for a privileged few. However, this vision depended on a settled world order—one in which India could rely on the United States to counter China's growing regional power,

*'India-China Clash: 20 Indian Troops Killed in Ladakh Fighting', 17 June 2020, https://www.bbc.com/news/world-asia-53061476. Accessed on 25 May 2026.

alongside industrialization and the deepening of the economic reforms that began in 1991.

Instead of following this path, India is now stuck. It sought to leapfrog into the Mumbai Consensus by focusing on services without sufficiently industrializing. China, however, pursued a different trajectory, completing market reforms and becoming a global supplier of industrial products. This created a unique dynamic: India imports far more from China than it exports, while relying on the US for export surpluses. With relations with the US turning sour, and sanctions increasingly masquerading as tariffs, the impact could amount to a significant share of GDP while simultaneously deepening dependence on China—where conflict, given competing interests, remains difficult to avoid.

The conflict with the US, however, threatens employment-intensive industries such as textiles, pharmaceuticals and agriculture. This potential compression of employment and exports could further constrain fiscal and monetary space. Yet, existing income compression across a large majority of the population implies that growth in services and other sectors must expand to provide fiscal room for income and food security support.

AI further worsens this already precarious situation. If current trends hold, many well-paying—perhaps even most—jobs may not exist in a few years. Startups, often celebrated as engines of opportunity, offer only partial solutions. They cater to narrow urban markets and globally connected consumers, and cannot alone generate the scale and depth of employment required. The consequences of this imbalance are profound, both socially and politically. If industrial growth and job creation lag behind population pressures, large segments of the workforce risk being underemployed, marginalized or pushed into precarious gig-based roles.

Economic exclusion can quickly translate into social tension, reduced upward mobility, and a pervasive sense of frustration among the youth. Preparing for this reality requires not only economic planning but also a cultural shift in how we value human work, technological literacy, and adaptability in a rapidly evolving employment landscape. The future of millions of Indians will depend not merely on AI or technology, but on our ability to align opportunity, skills and societal structures with the realities of the 21st century.

The history of migration offers a stark lesson. From the 1970s through the 1990s, Indians often moved abroad in search of education, economic opportunity and social mobility. Migration functioned as a safety valve, allowing talent to thrive where domestic systems could not provide opportunity. Today, however, the scale and nature of disruption are fundamentally different. The forces reshaping opportunity are simultaneously geographical, technological and political. As a result, the pathways that enabled past generations to succeed may no longer exist in the same form. Individuals, families and societies must rethink preparation, adaptability and skill-building in a compressed and competitive world.

Other nations may not have followed the same path as India, but Western societies too are experiencing rising costs of living and a sense of loss of the world they grew up in. Immigration, therefore, has become a source of anxiety even in the wealthiest and most industrially advanced countries. The pressures of technological disruption, combined with ageing populations and slowing growth, have transformed migration from a benign flow of talent into a perceived threat to economic security and social stability. Countries that once welcomed skilled workers as engines of innovation are increasingly tightening their borders. Policies are becoming more restrictive, visas more scrutinized and pathways to integration narrower.

This defensive posture is not limited to Europe or North America; it reflects a broader global recalibration in an environment where opportunity is scarce and the stakes are high. For those who once saw international mobility as a pathway out of domestic constraints, these closing doors are more than an inconvenience—they are a structural shift. Ambition alone is no longer sufficient; digital literacy, technological capability and the ability to navigate algorithmically mediated systems are becoming prerequisites for survival.

The consequences of such exclusion extend far beyond individual aspirations. When legal avenues to work and contribute are restricted, frustration and disenchantment grow, sometimes giving rise to alternative—and often risky—forms of livelihood. Cybercrime, digital fraud and even radicalization can become paths of least resistance for talented but constrained individuals. The very systems designed to regulate movement and opportunity may inadvertently generate vulnerabilities that threaten social stability both within and across borders, including within India itself.

How This Book Is Structured

The book is divided into two parts. The first part is introspective: we examine India and interrogate some of its foundational assumptions, highlighting why, despite confident narratives, India lacks sovereignty in digital systems and processes—that is, freedom of action. In the second part, we outline a roadmap and a set of actionable steps that could help reduce this constraint.

The first chapter examines a word that provokes intense frustration among Indians—KYC. We trace its origins to efforts to target Mexican and Colombian drug cartels by striking at their financial reserves, and its evolution into a broader sanctions architecture used by the West to strip the assets of geopolitical adversaries. By adopting these frameworks uncritically, and often implementing them poorly, India has curtailed the economic freedom of its population, undermined financial inclusion and eroded trust in institutions. At the same time, flawed design and execution have rendered these systems ineffective at curbing money laundering—their original objective under FATF guidelines. We also demonstrate how a combination of low-cost technological and policy interventions, including the use of AI and graph databases, could address many of these shortcomings.

The second chapter argues that industrialization, industrial policy and market reforms form a necessary troika for sovereignty. India did not fully pursue this path after liberalization, and therefore lacks the foundational capabilities required to lead in the AI era. Instead, it moved prematurely towards financialization within a predominantly services-led economy. In the absence of strong redistributive institutions, the gains from this model accrued to a narrow segment of the population. In contrast, China invested heavily in industrialization, and the surplus generated from its export-led growth created the societal and institutional base required to support advanced technological development, including the creation of cutting-edge LLMs.

The third chapter examines why customer service in Indian banks is persistently poor, tracing it to a structural dynamic in which a largely low-income population interacts with a banking system that is frequently used by the state as an instrument to achieve social and economic objectives, including financial inclusion. This dual compression makes it economically unviable for banks to serve retail customers at scale, leading them to prioritize wealthier clients and corporates. Poor customer service, we argue, is not incidental but systemic. Financial inclusion, in the absence of industrial inclusion, cannot generate income inclusion and instead risks trapping the system in cycles of rising household debt, increasing NPAs, and recurring credit booms and busts.

The fourth chapter explores how the concept of compression can help explain how LLMs function, and proposes a way of understanding their supply chain as an industrial system. We analyse how leadership in specific layers of this supply chain can enable the creation of foundational models—but only if a nation is an industrial, financial or research power. India, being none of these at scale, lacks the critical components required to build such models independently.

The fifth chapter examines how the interconnections between digital, financial and industrial supply chains can be exploited to create and manage parallel economies, including enabling forms of money laundering even on top of systems such as UPI.

The sixth chapter outlines a set of principles that could enable India to build leadership in disruptive technologies such as AI in a complex and volatile geopolitical environment. We describe these as the Five Principles of Digital Panchsheel: build for the world; embrace openness; govern with wisdom; retain agency and independence; and invest in commons and diverse algorithms. Together, these principles offer a pathway from uncertainty towards a future grounded in freedom.

The seventh chapter returns to the challenge of technology policymaking in an age of uncertainty. We show how policymakers often regulate the myth rather than the mathematics, and propose a framework for better governance: recognizing hidden architectures of power, shifting from regulation to stewardship, anticipating convergence, governing the unknown, resisting over-correction and ensuring broad-based representation.

In the concluding chapter, we argue that reality always asserts itself in the end. For India to navigate this era of uncertainty, self-criticism, reflection and introspection must become central. Sovereignty—defined here as freedom of action—can emerge only when the state enables such freedom for its citizens across economic, political, personal and intellectual domains. Only then can India aspire to emerge from this century as a consequential global power.

1

How KYC Got Twisted into an Extortion Scheme

It would be inaccurate to describe Mrs H as a digital native. She is something more than that—a digital optimist. It was her career in digital technology that shaped this outlook. She holds a master's degree in encryption and works in the same field for a well-regarded US company that specializes in building systems to store sensitive data securely. This background gave her a particular worldview: if a provider reliably promises that data is encrypted, there is little reason to suspect personal risk exposure. Unlike most non-technical users, she understood what PCI-DSS* meant. Her professional experience also gave her the confidence to choose what she believed were lower-risk fintech applications among the many UPI apps available.

Whether paying the newspaper *bhaiya* or her beautician, UPI via PhonePe was her default choice because it never failed—except on one day in July 2023. It was, at first, a small embarrassment not to have the ₹30 owed to the jasmine flower seller. Initially, she assumed it was a temporary technical glitch. But after a few more failed payments, realized that the linked bank account itself had been blocked, and that too without any prior notice.

A visit to the bank, which she had not entered in the last five

*The Payment Card Industry Data Security Standard (PCI DSS) is a global security standard developed to ensure the secure handling, processing, storage and transmission of cardholder data.

years, resulted in a near panic attack when the bank manager informed her that the block had been imposed following an order from cybercrime authorities, on suspicion that she might be involved in money laundering.

It was both disturbing and oddly inexplicable. She had used the account purely as a digital wallet, never keeping more than ₹25,000 in it and using it only for small, routine payments. The account itself had been in continuous use for 25 years, having been opened when she first left home to study engineering in another city. The bank manager explained that it was not any outgoing transaction that had triggered the freeze, but an incoming payment of ₹150.

This ₹150 had been sent by a cab driver in Hyderabad after her spouse paid him ₹500 in cash for a trip that cost ₹350. Lacking change, the driver offered to transfer the balance digitally, and her spouse provided her account details. Coincidentally—and with a certain tragic irony—the taxi ride itself had been taken to visit the Cyberabad Commissionerate to understand how law-enforcement agencies deal with cybercrime.

Does this mean that the cab driver was involved in money laundering? Most likely not. However, with the steady rise in cybercrime, it had become standard practice for law-enforcement agencies to trace outgoing transactions from accounts suspected of fraud across multiple levels—often up to six layers deep—and to issue lien markings or freeze orders along the way. Mrs H's account had simply entered this tracing chain because of a random incoming payment.

After several visits to the bank, she was eventually able to have her account unfrozen, but only after providing explanations, resubmitting identity and address documents, and assuring full cooperation with any future notices from law enforcement agencies. The experience left her with a bitter aftertaste and several deeply uncomfortable questions:

1. Given that her primary use case involves frequent small transactions, is she now expected to verify the antecedents of every individual who sends her money? And if so, how is she supposed to do that in practice?
2. How is it that she was never given any notice that her account had been flagged as suspicious, along with an opportunity to clarify?
3. In a system where bank details are no longer secret—where anyone can send money simply by knowing a linked phone number—does this mean that any malicious actor can implicate an individual by sending them a trivial amount?

Her response, instinctively, was to protect herself by reverting to cash and cards and reducing her use of UPI. But this was, in effect, a flight response to a stressful and disorienting experience.

Her case, however, pales in comparison to what was reported in Gujarat in 2024.* Police officials reportedly created ledgers of bank accounts, issued freeze orders on those accounts, and then demanded bribes to reverse them. Even as recently as 2025, similar patterns have appeared sporadically.† The modus operandi involved middlemen making small transactions into targeted accounts and then filing fraudulent cybercrime complaints. These accounts would then be frozen, after which the account holders would be approached and asked to pay a bribe to have them unfrozen—effectively converting the machinery of cybercrime

---

*Lavania, Sudeep, '3 Gujarat Cops Accused of Trying to Extort Money after Freezing Bank Accounts', *India Today*, 27 January 2024, https://www.indiatoday.in/india/story/gujarat-cops-accused-of-trying-to-extort-money-after-freezing-bank-accounts-case-filed-2494149-2024-01-27. Accessed on 25 May 2026.

†'Rajpipla Cybercrime Cell Cop Accused of Freezing Bank Accounts to Extort Money', *Gujarat Samachar*, 5 August 2025, https://english.gujaratsamachar.com/news/gujarat/rajpipla-cyber-cell-cop-accused-of-freezing-bank-accounts-to-extort-money. Accessed on 25 May 2026.

enforcement and KYC compliance into an organized extortion racket.*

The underlying cause that precipitated these incidents can be traced back at least 35 years—to the creation of the Financial Action Task Force (FATF) in 1989.

Drugs, Cartels, Illegal Exchanges and Forfeiture

FATF was established to address the growing problem of the global drug trade and the laundering of its proceeds.[7] The fact that G7 countries collectively identified this as a priority in 1989 illustrates how large the problem had become. The early 1980s were the era of Colombian drug cartels, which built vast empires spanning cultivation to retail distribution, generating so much cash that airplanes were required to transport US dollars back to Colombia. As George Jung famously recalled, 'It became more of a problem to count the money and stack it… It started to take the fun out of the whole thing.'†

These cartels funnelled cash back into the US financial system by walking into banks with suitcases full of currency and depositing them into accounts, often with the tacit cooperation of bank employees who facilitated the process with counting machines and structured deposits. Technically, this violated the Banking Secrecy Act (BSA) of the 1970s, which required record-keeping,[8] identity verification and reporting of transactions above US$10,000. In practice, however, these rules were widely ignored.[9] The BSA itself was challenged in the US Supreme Court, and even after being upheld, enforcement remained weak.

*'Narmada Cyber Crime Extortion Racket Exposes Police Role 1.5 Years after Junagadh Case', *Gujarat Samachar*, July 17 2025, https://english.gujaratsamachar.com/news/gujarat/narmada-cyber-crime-extortion-racket-exposes-police-role-1-5-years-after-junagadh-case. Accessed on 25 May 2026.

†Zill, Oriana, and Lowell Bergman, 'The Black Peso Money Laundering System', *Frontline*, PBS, https://www.pbs.org/wgbh/pages/frontline/shows/drugs/special/blackpeso.html. Accessed on 25 May 2026.

The situation changed dramatically when the US Treasury undertook currency-flow mapping[10] and identified that a disproportionate share of excess liquidity originated from banks in specific regions associated with drug proceeds. Banks were then pressured to report transactions more rigorously, enabling analysis, seizures and civil forfeiture. This, in turn, accelerated the growth of parallel currency-exchange markets.

While informal dollar–peso exchanges had always existed, the scale expanded significantly with drug money. It was estimated that nearly 20 per cent* of Colombia's money supply in 1979 was driven by drug-related inflows. The system operated roughly as follows:†

- A drug dealer who has accumulated cash in US dollars makes a deal with an (illegal) operator to exchange those dollars for pesos at a discounted rate, and then drops off the cash at a designated location. Once the cash is handed over, he has effectively offloaded his risk to the operator and expects an equivalent amount in pesos in Colombia.
- The operator then 'cuts' the cash into smaller amounts and, using runners he employs, deposits them into various US bank accounts under his control. These funds are then moved around, making them difficult to trace.
- Simultaneously, the operator's associates scout for legitimate businesses in Colombia that require US dollars to purchase goods from US corporations. He fixes an exchange rate (typically at a discount to the prevailing rate), collects pesos from these businesses, and delivers them to the drug dealer's associates in Colombia, thereby settling one side of the ledger.

---

*Keh, Douglas J., *Drug Money in a Changing World: Economic Reform and Criminal Finance*, no. 4, 1998, p. 3, fig. II, https://www.unodc.org/pdf/technical_series_1996-01-01_2.pdf. Accessed on 25 May 2026.

†'Interview: Mike Wald', *Frontline*, PBS, https://www.pbs.org/wgbh/pages/frontline/shows/drugs/interviews/wald.html. Accessed on 25 May 2026.

- He then wires the required US dollars to the bank accounts of US corporations for the purchase of goods, thereby settling the other side of the ledger, while making a profit on the exchange-rate differential on both sides of the transaction.

The narcotics trade not only enriched the cartels but also increased the sales of US Fortune 500 companies, as traffickers laundered money into purchases such as refrigerators[11] manufactured by General Electric, among others. Not surprisingly, this created a trade surplus for the US with Colombia, to the extent of US$4 billion in the 1990s. The money-laundering industry, spurred by the drug trade, thus functioned as an effective wealth-distribution mechanism—from US consumers to US corporations—with drug cartels acting as intermediaries and capturing a significant share of the transfer.

Consumers addicted to drugs spent increasing portions of their income on narcotics, and the resulting income accruing to drug dealers expanded the money supply in Colombia.* This, in turn, increased demand for US-produced goods and drove higher sales for Fortune 500 companies, contributing to a net balance-of-payments surplus.[12] In the long term, however, this dynamic also damaged the Colombian economy. The ease with which goods such as GE refrigerators could be purchased through discounted, laundered dollars in the shadow market reduced incentives to build domestic manufacturing capacity.[13]

Law-enforcement agencies, therefore, began a series of actions against Fortune 500 companies. These included seizing funds from their bank accounts,† where there were reasonable grounds to believe that they had wilfully ignored—or blinded themselves to—

---

*Keh, Douglas J., *Drug Money in a Changing World: Economic Reform and Criminal Finance*, no. 4, 1998, p. 3, fig. II, https://www.unodc.org/pdf/technical_series_1996-01-01_2.pdf. Accessed on 25 May 2026.

†This is called civil forfeiture, which is a legal mechanism that permits law enforcement agencies to seize assets suspected of being connected to criminal activity. In the US, it was significantly expanded under the Comprehensive Crime Control Act of 1984, enacted during the presidency of Ronald Reagan.

the possibility that some of their customers were engaged in money laundering. Companies were also required to enter into agreements mandating preventive measures, such as conducting due diligence on customers and distributors in other countries.

Forfeiture struck at the core of the drug trade, because the business was ultimately about money, not drugs. When that money was lost, it was often unclear to the cartels whether it had been confiscated by law enforcement or stolen, leading to distrust, internal violence and a contraction of operations. However, this approach also produced pernicious side effects. The war on drugs required funding, and the Crime Control Act allowed law-enforcement agencies to retain a portion of seized assets—up to 80 per cent.* Predictably, forfeitures and seizures increased dramatically, rising from US$25 million in 1985 to US$644 million in 1991.

Law enforcement, in effect, risked being incentivized into overreach. Unlike criminal forfeiture, which requires a trial and conviction, civil forfeiture requires only suspicion and no evidentiary standard of proof, and therefore succeeds more frequently. This creates strong incentives to demonstrate performance through a single metric—the monetary value of assets seized—which may bear little relation to actual criminal activity.

Forfeiture thus elevates a legal fiction into operational reality.[14] There may be no identifiable criminal to prosecute, no trial to conduct, yet there is presumed to be a crime committed by an instrument—money or an asset of monetary value. That instrument is therefore seized, detained, 'tried' and effectively convicted in a single step. Since the instrument cannot defend itself (unless the amounts are substantial enough to justify a legal challenge), the conviction rate remains high, reinforcing the system.

To recap, the freezing of Mrs H's bank account—based on

---

*Dunn, Kyla, 'Reining in Forfeiture: Common Sense Reform in the War on Drugs', *Frontline*, PBS, https://www.pbs.org/wgbh/pages/frontline/shows/drugs/special/forfeiture.html. Accessed on 25 May 2026.

suspicion that it may have been an instrument used in laundering a sum as small as ₹300—can be traced to the recommendations of the FATF. These recommendations originated in America's war on drugs and gave rise to a new class of legal mechanisms that treat assets as if they were persons, subject to action without due process. The conceptual roots of this approach can be traced back to the 1827 case *The Palmyra*, in which the US Supreme Court endorsed the legal fiction of a 'guilty object'.*

The Strange Evolution of KYC in India

If one wished to open a bank account in the early 1990s, there were no lengthy forms to fill out and no identity proofs to submit. A handshake, a name entered into a ledger, and a letter of introduction from an existing customer or a trusted member of the community—that was often sufficient. Trust was both personal and societal, built on reputation and relationships. The reliance on introducers was not because bankers were inherently suspicious of customers or treated them as potential money launderers, but because it served as an important safeguard under Section 131 of the Negotiable Instruments Act.[15]

This provision arose because, in the absence of digital records, settlement delays created opportunities for fraud. Until 1988, nothing effectively prevented an account holder from issuing cheques exceeding their available balance, prior to the insertion of Chapter XVII.† Cheque alterations were also common, and bankers were therefore required to exercise due diligence not only in validating the instrument itself but also in assessing the credibility of the account holder. Self-drawn cheques were expected

---

* *The Palmyra*, 25 U.S. (12 Wheat.) 1 (1827), *Reports of the Decisions in the Supreme Court of the United States*, January 1827, pp. 14–15, https://tile.loc.gov/storage-services/service/ll/usrep/usrep025/usrep025001/usrep025001.pdf. Accessed on 25 May 2026.

†*Banking, Public Financial Institutions and Negotiable Instruments Laws (Amendment) Act, 1988* (Act No. 66 of 1988), https://www.casemine.com/act/in/5a979d974a93263ca60b70ce#1001_1_FN0001. Accessed on 25 May 2026.

to be presented personally; when submitted through third parties, they required a letter of authorization.

Relationships and personal references from trusted, long-standing customers thus formed the basis of 'knowing your customer'. The presence or absence of formal identity documents was not decisive in opening an account. This is evident from exemption clauses that explicitly stated that photographs need not be insisted upon in certain cases—for instance, in new savings accounts where cheque facilities were not provided.[16]

The introducer system therefore functioned as a safeguard against conversion risk that could be transferred to the bank when a customer submitted a negotiable instrument (such as a cheque or demand draft) for collection. At that point, the banker acted as the customer's agent and could be held liable to the instrument's rightful owner. Case law reinforced this principle. In *The Bharat Bank Ltd. vs Khishinchand Chellaram* (1953),* for example, the bank was held liable for cheque fraud because it had failed to make adequate enquiries at the time of account opening.

The legal foundation of the introducer model was eventually displaced by the Forty Recommendations of FATF, issued in 1990.[17] Recommendations 12, 13 and 14 called for client identification based on official or otherwise reliable documentation. India, however, continued to follow the introducer model until the enactment of the Prevention of Money Laundering Act (PMLA),[18] which required financial institutions to verify and maintain records of all clients in a prescribed manner.

Document-based identification became the primary method of opening bank accounts after 2005, when the PMLA rules were notified.[19] However, an important complication arose in the design of client due diligence (CDD), which can be conducted as specified in Rule 9.[20] It permits reliance on a 'trustworthy third party',

---

**The Bharat Bank Ltd., Madras v. Khishinchand Chellaram*, O.S.A. No. 130 of 1950 (Madras High Court, 25 September 1953).

which may even be located outside India, provided it fulfils the obligations imposed under the Act. This creates a set of structurally incompatible obligations:

- The financial institution remains accountable for the integrity of the account-opening process (Rule 9, Clause 2(e)).
- Yes, its judgement is effectively dependent on the trustworthiness of a third party responsible for maintaining and issuing identification documents and data.

This design fails a fundamental test of accountability. One party (the financial institution) is accountable but not directly responsible for the underlying data, while the third party is responsible for generating and maintaining the data but is not fully accountable for its downstream use.

Consider, for instance, the Central KYC (CKYC) registry,[21] which is authorized to maintain a central database of identification records. CKYC does not independently verify or validate the information at the point of creation; instead, it relies on reporting entities to do so.[22] As specified, 'the reporting entity which performed the last KYC verification or submitted updated information shall be responsible for verifying the authenticity of the identity or address of the client.' This introduces yet another layer of delegation, further weakening the reliability of the identity-verification process.

But how does any reporting entity verify the truthfulness or accuracy of identity documents, including addresses? To answer that, one must examine the Aadhaar system and the broader history of identity-document regimes that preceded it.

When Is a Human *a* Human?

Human beings never exist in isolation, even when they renounce the world. They are always embedded in a symbiotic relationship with other humans. Hence, human identity is not merely individual;

it is also derived from the relationships individuals have with other human groups—such as friends, family, tribe, caste, village, kingdom (in earlier eras), and the nation-state (in the modern era).

Across the world, therefore, names were not purely individualistic but often reflected lineage and community. They included the name of the parent and, in some cases, even the name of the village of origin or that of grandparents. Tribe or caste identifiers were appended, and in many societies, women adopted their husbands' names. Identity documents, therefore, were historically issued by local community leaders and were accepted as valid within those contexts. For instance, consider the list of documents accepted by telecom providers[23] in India: it includes a certificate of identity with a photograph issued by a village panchayat head, which can also serve as proof of address if it includes the address.

There are, however, recurring problems when identity documents are issued by local authorities. First, such documents may not be accepted outside the local jurisdiction. Second, the system can be gamed, as local authorities may be bribed to issue false certificates. Broadly, these problems are addressed through:

- Creating a legal framework and establishing identity authorities* that issue documents valid across wider jurisdictions.
- Introducing de-duplication and fraud checks on documents issued by such authorities.

De-duplication, however, belongs to a class of problems often described as 'wicked problems', because there are no easy solutions at scale. Given the density of populations, any biographic data chosen will inevitably produce collisions.†

---

*Passports are one such example: countries collectively recognize them as time-bound identity documents issued for international travel. Citizenship IDs are another, serving as widely accepted proof of identity and address within national boundaries.

†Collision is a technical term indicating that a field or identifier may produce duplicate values—that is, it is not guaranteed to be unique and may therefore result in overlaps.

In the case of Aadhaar, the prevailing hypothesis for solving the de-duplication problem rested on three constructs:

1. Biometrics are unique to every human being (for example, fingerprints).
2. Even if individual biometric markers are not unique, a combination of them (such as iris scans, fingerprints and palm prints) would be unique when taken together.
3. Biometric information, once captured, would not be subject to loss or tampering.

All three assumptions have proved to be flawed—not only in implementation, but also in theory. To understand why, consider evidence from the field. According to a report in *The Times of India* (18 March 2023)* the Delhi Police found that:

- The de-duplication algorithm treated all 10 fingerprints and the iris as a single package, rather than as independent variables. Fraudsters were therefore able to generate multiple Aadhaar numbers for the same individual by mixing and matching fingerprints from different individuals while using the same photograph.
- Facial biometrics were not incorporated into the de-duplication process; otherwise, the same individual using the same photograph could not have obtained multiple identities.
- Iris scanners could be fooled by capturing images from high-resolution colour photographs, as there were no liveness checks.
- Fingerprint scanners could similarly be fooled using silicone fingerprints, again due to the absence of liveness checks.

*Jha, Rajshekhar, 'How Crooks Are Exploiting Gaps in Aadhaar System in Delhi', *The Times of India*, 18 March 2023, https://timesofindia.indiatimes.com/city/delhi/how-crooks-are-exploiting-gaps-in-aadhaar-system-in-delhi/articleshow/98744284.cms. Accessed on 25 May 2026.

This was not an isolated incident. Such vulnerabilities were first reported as early as 11 September 2017,* and it is likely that government agencies were aware of them even earlier.

Crucially, the fraudsters were not external actors but part of the enrolment ecosystem itself, as both the 2017 and 2023 incidents demonstrate. This was further corroborated by UIDAI when it barred Common Service Centre (CSC)[24] from participating in enrolment for a period of two years, citing widespread corruption driven by a perverse incentive structure. Operators were compensated only when enrolments resulted in the generation of Aadhaar numbers, incentivizing them to exploit vulnerabilities—of which there were many across theory, design and implementation. The most fundamental flaw lay in the assumption that biometrics are inseparable† from the human body.

In effect, while UIDAI began with the premise that an identity document backed by biometrics would be more reliable than one without,‡ the reality proved otherwise. In practice, a person became 'verified' not by intrinsic identity but by the assertion of an enrolment operator, validated by a flawed de-duplication algorithm—one that fraudsters were able to exploit extensively over a prolonged period.

Digital KYC Meets Physical Reality

KYC in India in the post-Aadhaar era can, in effect, be described as an information exchange within a whisper network—where not all

---

*Shekhar, Shashankh, 'UP STF Busts Gang Which Hacked Secure Source Code to Crack Aadhaar, Issued Fake Biometric Cards', *India Today*, 11 September 2017, http://indiatoday.intoday.in/story/aadhaar-card-fake-biometric-id-up-stf-uidai/1/1045133.html . Accessed on 25 May 2026.

†Severability refers to the ability to separate one element from another. A foundational assumption in Aadhaar is that biometrics are inseparable from the individual. However, this assumption is theoretically contestable, as biometric attributes—such as fingerprints—can be altered through medical or cosmetic procedures.

‡The numerous orders issued from 2014 onwards mandating the linkage of Aadhaar with other identity documents, such as PAN, were explicitly based on this premise.

whispers are trustworthy, and therefore the burden of due diligence ultimately falls on the receiving entity, often at additional cost.

Consider Aadhaar-based electronic KYC. It provides demographic details (name, date of birth, photograph) and address information. Yet there is no guarantee that any of this information is accurate, as the preceding discussion illustrates. There is typically no physical verification of address, nor independent validation of identity, unlike in the case of passport issuance. If the individual is truthful, the data will be accurate; if not, it will not be. The most appropriate way to understand Aadhaar, therefore, is as a form of self-certified identity.

Until recently, however, regulators and regulated entities did not fully acknowledge the risks arising from data-quality issues in CKYC and Aadhaar databases. Customer onboarding was permitted through e-KYC* alone, largely because it was inexpensive. This contributed to an increase in fraud, non-performing assets arising from 'skip' cases† and the proliferation of money mules during the pandemic period.

Only in April 2023,‡ did the regulator classify accounts opened through faceless modes—such as Aadhaar OTP, DigiLocker, CKYC and offline Aadhaar XML—as high-risk, and mandate either physical verification or video-based customer identification (V-CIP).

The timing of this regulatory intervention is noteworthy. It followed closely on the March 2023 reporting by the Delhi

---

*Electronic KYC using Aadhaar can be carried out through several methods: (1) OTP-based authentication, where the holder is verified using a one-time password sent to their registered mobile number; (2) offline verification through a shared XML file; (3) submission of a photocopy of the Aadhaar document; (4) use of a CKYC identifier; and (5) sharing documents via DigiLocker.

†A 'skip case' refers to a borrower who, after taking a loan, becomes untraceable and fails to repay.

‡Reserve Bank of India, KYC *Master Direction*, November 2024, p. 43, para. 40, https://www.rbi.org.in/commonman/Upload/English/Notification/PDFs/MD18KYCF6E92C82E1E1419D87323E3869BC9F13.pdf. Accessed on 25 May 2026.

Police* on Aadhaar data-quality issues. The industry response was predictable: concerns about increased costs,[25] typically ranging from ₹30 to ₹100 per customer, compared to near-zero costs when relying on CKYC or DigiLocker.

Does this imply that financial institutions are indifferent to KYC or money laundering risks? Not necessarily. Like regulators, they appear to have overestimated the extent to which digital KYC could substitute for the inherently physical process of customer due diligence—an assumption reinforced by official communications.

For instance, a PIB release dated 8 September 2023 suggested that KYC costs had declined from US$0.12 to US$0.06 (approximately ₹10 to ₹5). The Economic Survey presented a far more dramatic reduction—from US$12 to US$0.06 (₹1,000 to ₹5)†—while industry figures such as Nandan Nilekani‡ suggested an even greater drop, from US$23 to US$0.05 (₹2,000 to ₹5). Clearly, only one of these estimates can approximate reality, and even that may not reflect actual ground conditions.

To recap, Indian banks have moved from an introducer-based, relationship-driven model of account opening in the 1990s, to document-based identification under FATF-influenced regulations in 2005, to faceless, paperless, presence-less digital KYC around 2020, and back to physical or video-based verification by 2024 under regulatory direction.

---

*Jha, Rajshekhar, 'How Crooks Are Exploiting Gaps in Aadhaar System in Delhi', *The Times of India*, 18 March 2023, https://timesofindia.indiatimes.com/city/delhi/how-crooks-are-exploiting-gaps-in-aadhaar-system-in-delhi/articleshow/98744284.cms. Accessed on 25 May 2026.

†Kawale, Ajinkya, 'Economic Survey: DPI Brings Down KYC Costs to Rs 6 from about Rs 1,000', *Business Standard*, 22 July 2024, https://www.business-standard.com/economy/news/economic-survey-dpi-brings-down-kyc-costs-to-rs-6-from-about-rs-1-000-124072201118_1.html. Accessed on 25 May 2026.

‡Rao, Rukmini, 'Nandan Nilekani Outlines a Blueprint for India to Unlock $8 Trillion Economy by 2035', *Fortune India*, 13 March 2025, https://www.fortuneindia.com/business-news/nandan-nilekani-outlines-blueprint-for-india-to-unlock-8-trillion-economy-by-2035/121129. Accessed on 25 May 2026.

In the process, however, they temporarily debanked a significant proportion of customers by freezing accounts pending KYC updates, while simultaneously exposing them to fraud. Fraudsters exploited the coercive threat of account freezes to extract money—amounting to crores—from unsuspecting account holders.

KYC Is Broken because the Regulator and the Government Broke It

The regulatory architecture in India for financial institutions has been unambiguous and stable for close to two decades, and broadly follows a three-tier structure:

1. Parliament passes legislation, which often contains provisions enabling the executive to frame rules for implementation. For instance, in the case of PMLA, Section 73[26] allows the Central Government to notify rules for implementing the Act.
2. Rules are then notified under these powers for operationalizing the law. In the case of PMLA, although Parliament passed the legislation in 2002, the rules were notified only in 2005.
3. The sectoral regulator (for example, the RBI) then issues directions containing specific guidance for regulated entities to comply with the law and its operational rules. In the case of PMLA, the RBI issued the Master Directions on KYC in 2016,* consolidating and superseding various circulars issued since 1976.

The Master Directions, however, are not optional guidance for regulated entities. The Department of Banking Supervision (DoBS)[27]

*Reserve Bank of India, *Master Direction – Know Your Customer (KYC) Direction*, 2016, Master Direction DBR.AML.BC.No.81/14.01.001/2015-16, 25 February 2016 (updated as on 14 August 2025), https://www.rbi.org.in/commonman/English/scripts/notification.aspx?id=2607. Accessed on 25 May 2026.

within the RBI conducts periodic on-site inspections,* along with continuous off-site surveillance between inspections. Supervisory concerns are routinely escalated to the senior management of banks, accompanied by expectations of immediate rectification and time-bound action plans that are closely monitored.

One of the primary reasons banks freeze accounts for want of KYC updates lies in the Master Directions on KYC, which require banks to classify accounts into three risk categories—low, medium and high—with periodic updates mandated at intervals of 10, 8 and 2 years respectively (depending on risk level).[28] While the Directions are largely silent on the specific actions to be taken when customers fail to update their KYC, it is not difficult to infer the likely response of banks when questioned by the regulator. A supposed conversation:

Regulator: Can you provide a dataset containing the following fields for all accounts—name, account number, IFSC, date of opening, risk category, and last KYC update date?

Chief Risk Officer (CRO): Here is the sheet.

Regulator: Why have 20 per cent of these accounts not completed KYC updates in accordance with Section 37?

CRO: That is a gap. We will address soon.

Regulator: This is a serious gap. I want an action plan to close this within 30 days.

CRO: Yes, done.

What follows internally is predictable: senior management pushes compliance teams for rapid closure; compliance teams push branch managers; branch managers, constrained by time and resources and

*Inspection is a more rigorous and exacting process compared to the relatively broader and more lenient concept of an audit.

unable to reach thousands of customers, resort to freezing accounts—hoping that customers will either visit the branch or complete online KYC, thereby satisfying both internal management and the regulator.

Is such a chain of actions plausible? Field reports suggest that it is. For instance, *Moneylife* reports: 'Banks, in turn, said that RBI inspectors harass them if they do not take the extreme step of freezing accounts if KYC was not updated.'* But does such action have any legal basis in the Master Directions or in legislation passed by Parliament?

The RBI's own report states: 'Freezing of customer deposit accounts for non-submission of KYC was reported to be a practice in many REs, even though the regulation does not provide for it.'† If so, what then formed the basis for widespread account freezes? An RTI filed by the Moneylife Foundation in January 2021 offers a clue: 'It may also be noted that provisions on partial freezing were applicable prior to the amendment on April 20, 2018, to the Master Directions on KYC dated February 2016.'‡

This leads to a further question: what triggered the RBI to amend the Directions on 20 April 2018 and remove the provision for partial freezing?

The answer lies in the Aadhaar litigation. The PMLA rules mandating linking of Aadhaar to bank accounts—failing which

---

*Basu, Debashis, 'KYC: Banks Don't Have the Right to Freeze Accounts. It's Illegal', *Moneylife*, 23 June 2023, https://www.moneylife.in/article/kyc-banks-dont-have-the-right-to-freeze-accounts-its-illegal/71135.html. Accessed on 25 May 2026.

†Reserve Bank of India Consumer Education and Protection Department, *Report of the Committee for Review of Customer Service Standards in RBI Regulated Entities*, 24 April 2023, p. 44, para. 4.4.2, https://fidcindia.org.in/wp-content/uploads/2023/06/RBI-REPORT-OF-THE-COMMITTEE-FOR-REVIEW-OF-CUSTOMER-SERVICE-STANDARDS-IN-RBI-REGULATED-ENTITIES-05-06-23.pdf. Accessed on 25 May 2026.

‡Dalal, Sucheta, 'Should Banks Be Permitted To Freeze a Running Bank Account, for KYC?', *Moneylife*, 27 January 2021, https://www.moneylife.in/article/should-banks-be-permitted-to-freeze-a-running-bank-account-for-kyc/62771.html. Accessed on 25 May 2026.

accounts could be frozen—were challenged before the Supreme Court. While the Court reserved its judgement on 10 May 2018, it was evident from the hearings that the bench was sympathetic to arguments concerning 'financial death' caused by account freezes. In its final judgement in September 2018, the Court struck down the mandatory linking of Aadhaar to bank accounts, except where required for receiving state subsidies and benefits.

Therefore, the freezing of bank accounts for failure to complete KYC updates has not only lacked regulatory backing since April 2018, but has arguably been unconstitutional since September 2018. Yet, the practice continued widely, without meaningful opposition from the RBI—a central observation of the Kanungo Report.

While it may not be possible to conclusively determine the motivations of the regulator in nudging entities towards such practices, the consequences were predictable. The fear of 'financial death'—loss of access to one's own bank account—renders individuals highly vulnerable to exploitation.

Fraudsters have leveraged this fear through multiple methods, but the underlying objective remains the same: to obtain access to customers' banking credentials through coercion, deception or a combination of both. Coercion often involves inducing customers to share OTPs sent to their registered mobile numbers or disclose debit card details. Such schemes do not require sophisticated techniques—only an ability to mimic the behavioural patterns of institutional authority.

As Sucheta Dalal has documented, reflecting on her own experience with business banking:*

> One official at the Bank refused to touch the resubmitted papers or discuss the issue, while the branch-head used to summon our accounts manager and give him vague oral instructions, refusing to put things in writing. If this happens

*Ibid.

to a media organisation (which usually get a response), one can only imagine what happens to customers with no access.

The monetary loss suffered by customers due to KYC-related account freezes has not been systematically tracked by the regulator. However, data from the Indian Cyber Crime Coordination Centre (I4C), which tracks fraud by modus operandi, allows for a conservative estimate.

Until 2023, I4C reported a net loss of ₹8,586 crore.* The figure is partial, and subsequent reports† indicate that in 2024, losses rose to approximately ₹60 crore per day—equivalent to ₹21,900 crore annually. This yields a combined reported loss of ₹30,486 crore. If one assumes a 60:40‡ split between international and domestic fraud operations, ₹12,194 crore can be attributed to domestic networks.§ If 40 per cent of this relates to KYC and refund frauds, and assuming an even split between the two, a conservative lower bound for KYC-related fraud emerges at approximately ₹2,400 crore.

Estimating an upper bound is more difficult, given the absence of granular data from banks and regulators. However, a proxy can be derived from funds lying idle in Jan Dhan Yojana (JDY) accounts.

A bit of context on why this is a good proxy is necessary. JDY accounts were opened en masse, often with a single document (Aadhaar) after 2014, to transfer money to beneficiaries via Direct Benefit Transfer.[29] However, more than 67 per cent of the population

---

*Standing Committee on Communications and Information Technology (2023–24), *Digital Payment and Online Security Measures for Data Protection: Fifty-Fourth Report*, Ministry of Electronics and Information Technology, Lok Sabha Secretariat, February 2024, p. 9, https://sansad.in/getFile/lsscommittee/Communications%20and%20Information%20Technology/17_Communications_and_Information_Technology_54.pdf?source=loksabhadocs. Accessed on 25 May 2026.

†Indian Cyber Crime Coordination Centre (I4C), *Recent Cyber-crime Trends in Financial Frauds*, 12 August 2024, slide 2, https://ciet.ncert.gov.in/storage/app/public/files/17/Workshop_and_training_files/fsch/12aug.pdf. Accessed on 25 May 2026.

‡This is an arbitrary assumption, but a reasonable one to adopt.

§Indian Cyber Crime Coordination Centre (I4C), *Recent Cyber-crime Trends in Financial Frauds*, 12 August 2024, slide 5, https://ciet.ncert.gov.in/storage/app/public/files/17/Workshop_and_training_files/fsch/12aug.pdf. Accessed on 25 May 2026.

opened bank accounts using Aadhaar as a paper-based ID, driven by the target-based approach adopted by the government. Around 97 per cent of these accounts were opened by public sector banks, which carried out account opening at the district and state levels under the coordination of State Level Bankers' Committees.

Top-down mandates and pressure from both the government and the regulator meant that, for banks, financial inclusion became a cost-centre activity and was not viable as a commercial model. The primary objective of the exercise was attribution of credit to the government for delivering direct benefit transfers.[30] For beneficiaries, too, the conditions were far from ideal. The most common reasons cited for inactive accounts—accounts opened in their name but with no subsequent debit or credit activity—include distance from financial institutions, lack of trust in the institution, lack of need for the account, and insufficient funds.

Why would beneficiaries lack trust in the institution? There may be several reasons, but they all converge on one material outcome: unlike cash, money in such accounts is not always accessible when needed. This could be because the institution is physically inaccessible, does not treat them well, or simply does not allow them to withdraw money. The first two reasons relate to commercial viability and service delivery; the third can only arise from account-level restrictions, such as freezes.

The Global Findex database* indicates that approximately 5 per cent of respondents cite lack of trust as the sole reason for maintaining an inactive account. This can therefore be used as a reasonable proxy for estimating the upper bound of the monetary impact of KYC-related restrictions.

---

*Demirgüç-Kunt, Asli, Leora Klapper, Dorothe Singer, and Saniya Ansar, *The Global Findex Database 2021: Financial Inclusion, Digital Payments, and Resilience in the Age of COVID-19*, World Bank, 2022, p. 113, fig. 2.5.3, https://openknowledge.worldbank.org/server/api/core/bitstreams/1d7f4edc-2b29-5f9a-b8aa-6e980a32d8ff/content. Accessed on 25 May 2026.

As of December 2024, approximately ₹14,750 crore* lay unused in dormant JDY accounts. If even 5 per cent of this is inaccessible due to KYC-related freezes, the impact amounts to at least ₹737 crore. This figure is almost certainly an underestimate when extended to the full universe of bank accounts affected by KYC-related restrictions. Even without such expansion, the monetary loss to the population attributable to faulty KYC practices is at least ₹3,137 crore.

However, unlike large banking frauds—such as the PMC Bank case, where losses of ₹4,000 crore had a clear focal point due to the identifiable institution—the losses here are distributed across millions of individuals. As a result, they do not attract comparable public attention, reflecting an asymmetry of power and visibility.

When the regulator and the government themselves erode public trust in the banking system by failing to prevent such practices, the question that remains is stark: what recourse does the ordinary citizen have?

Root Cause: Infinite Delegation Caused by a Sanctions Regime

The primary reason KYC remains broken lies in what may be described as infinite delegation—where entities responsible for providing clear direction on policy design and implementation continue to push responsibility downwards to other entities that are less equipped to understand the broader purpose of KYC, which ultimately originates in the FATF-driven sanctions regime.

The sanctions regime is fundamentally about penalizing individuals suspected of criminal activity. It does not require proof in a court of law that a crime has been committed; suspicion is sufficient to trigger action. Once suspicion arises that an individual

---

*'11.30 Crore PM JanDhan Accounts Are Inoperative with Rs14,750.27 Crore Balance: Govt', *Moneylife*, 13 December 2024, https://www.moneylife.in/article/1130-crore-pm-jandhan-accounts-are-inoperative-with-rs1475027-crore-balance-govt/75846.html. Accessed on 25 May 2026.

may have acquired illicit wealth, the system proceeds to freeze or attach that wealth until innocence or guilt is determined. In practice, however, the process itself becomes the punishment. Even when individuals are eventually found to be innocent, they are rarely compensated, as pursuing remedy is often prohibitively costly in terms of time and resources.

The sanctions regime thus operates as a legal fiction—it purports to offer due process, but in reality imposes punishment based on suspicion alone.

Infinite delegation, in turn, provides an escape valve for policymakers. They can argue that the actual harm is not caused by policy design but by the actions of entities further down the chain. To understand the depth of this delegation model, the following diagram would be useful:

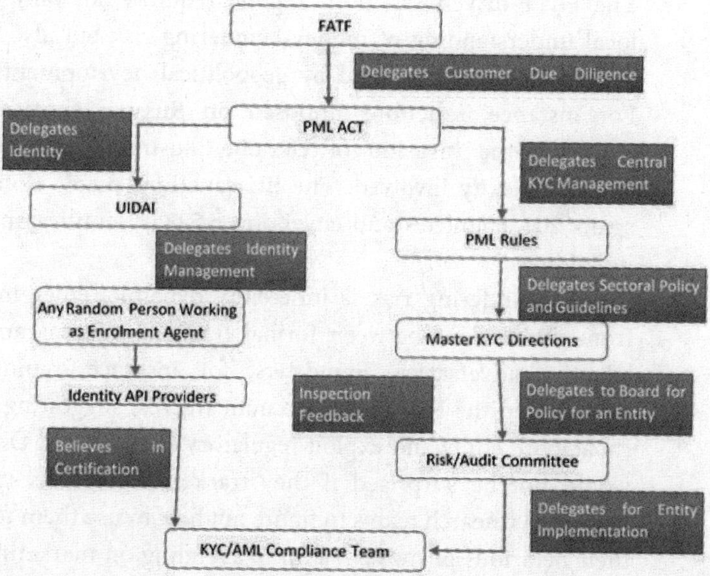

Operational teams within any given entity are at least six levels removed from those who frame the FATF recommendations and are at least three levels away from the supervisory teams of regulators. These teams typically lack agency in shaping policy implementation and are instead evaluated on compliance metrics—how well they perform in inspections and how effectively they avoid adverse observations.

As a result, when regulations are prescriptive, they have little choice but to follow them to the letter, even when they possess ground-level knowledge that such implementation may be ineffective or cause unnecessary hardship to customers.

While, in theory, regulatory directions require banks to adopt a risk-based approach—categorizing accounts as low, medium or high risk based on internal assessments—in practice, institutions and their anti-money-laundering (AML) teams are poorly positioned to exercise such judgement for several reasons:

- The FATF-driven sanctions regime requires not only a local understanding of money-laundering risk but also a global perspective shaped by geopolitical developments. For instance, sanctions imposed on Russian entities[31] following the invasion of Ukraine illustrate the scale and complexity involved. The EU sanctions map[32] alone spans 20+ countries and covers over 5,000[33] entities and individuals.
- Money-laundering risk is inherently dynamic, emerging from interactions between formal financial systems and adaptive adversaries. Fraudsters, for instance, rapidly weaponized the KYC-based account freezes, suggesting a capacity to study and exploit regulatory frameworks. One would not be surprised if they tracked regulations via dedicated research teams to figure out how to use them for their nefarious purposes—akin to spending on marketing or product research and development. AML teams, by contrast, are cost centres and hence, by their very nature, lack incentives to innovate in methods of lawbreaking as

their livelihood is not dependent on it.

- Thus, their idea of ML risk is confined to compliance with a checklist given to them by regulators and following it to the letter. Hence, their risk categorization becomes aligned with peer practices and regulatory expectations.*

Way Forward: Rethinking Money-Laundering Risk

If infinite delegation is the problem, where should it be arrested? The most appropriate level is the sectoral regulatory level because they have the best advantage when it comes to their sector. Delegating further upwards is ineffective, as laws such as the PMLA operate horizontally across sectors. Regulated entities are then best positioned to do what they do best—implement and enforce rules—without being required to innovate or interpret them.

FATF guidelines themselves allow for flexibility. For instance, the updated recommendations (February 2025) note that:†

> When assessing the money laundering and terrorist financing risks relating to types of customers, countries or geographic areas, and particular products, services, transactions or delivery channels risk, a financial institution should take into account risk variables relating to those risk categories. These variables, either singly or in combination, may increase or decrease the potential risk posed, thus impacting the appropriate level of CDD measures. Examples of such variables include:

---

*The RBI's KYC Directions are prescriptive, requiring periodic re-KYC at intervals of 2, 8 and 10 years based on high-, medium- and low-risk classifications. This reflects a static risk model: even where a regulated entity has a more sophisticated risk assessment framework, it must adhere to the regulator's prescribed timelines.

†Financial Action Task Force (FATF), *International Standards on Combating Money Laundering and the Financing of Terrorism & Proliferation: The FATF Recommendations* (2012–2025), FATF, Paris, p.72, para. 19, https://www.fatf-gafi.org/content/dam/fatf-gafi/recommendations/FATF%20Recommendations%202012.pdf.coredownload.inline.pdf. Accessed on 25 May 2026.

- The purpose of an account or relationship.
- The level of assets to be deposited by a customer or the size of transactions undertaken.
- The regularity or duration of the business relationship.

On simplified customer due diligence (CDD), the guidance states that:*

> Where the risks of money laundering or terrorist financing are lower, financial institutions should be allowed and encouraged to implement simplified CDD measures, which should take into account the nature of the lower risk. The simplified measures should be proportionate to the lower risk factors (e.g. the simplified measures could relate to customer acceptance measures, aspects of ongoing monitoring, or both). Examples of possible measures are:
>
> - Verifying the identity of the customer and the beneficial owner after the establishment of the business relationship (e.g. if account transactions rise above a defined monetary threshold).
> - Reducing the frequency of customer identification updates.
> - Reducing the degree of on-going monitoring and scrutinising transactions, based on a reasonable monetary threshold.
> - Not collecting specific information or carrying out specific measures to understand the purpose and intended nature of the business relationship, but inferring the purpose and nature from the type of transactions or business relationship established.

Taken together, these principles provide a basis for rethinking money-laundering risk at the sectoral level.

*Ibid, p. 73, para. 21.

Longevity of the Relationship Matters

It is well established that in India, most population movement occurs within states[34] rather than across them. Migration is often driven by employment (for men) or marriage (for women), and only about one in three individuals is a migrant. This suggests that, for a significant portion of the population, once identity is established, for at least 67 per cent of the population addresses would not vary after a period, as they have stabilized in a particular location. This also implies long-term relationships with banks—provided service-related concerns,* particularly among younger customers† who often report dissatisfaction with banking experiences, are addressed.

In such a context, there is little justification for repeatedly subjecting older or long-standing customers to KYC updates based on rigid, calendar-driven cycles (2, 8 or 10 years), irrespective of actual risk.

The primary countervailing risk arises with age: individuals may delegate account operation to someone younger as their physical and cognitive faculties start declining. This creates the possibility of divergence between ownership and control. However, this risk can easily be mitigated through transaction-based monitoring—such as income-profile models that detect shifts in behaviour and trigger risk reclassification when patterns deviate significantly.

To put it simply, what matters over time is not static identity verification, but changes in transaction patterns and control over the account.

Account-Transaction Profiling via Graph Neural Networks

Consider a recent case study in cybercrime involving money

---

*Nagabhushanam, Manasa, *A Study of Customer Service Quality in India*, https://iibf.org.in/documents/reseach-report/Report-26.pdf. Accessed on 25 May 2026.

†Ibid, p. 361, ch. 9. The study notes that the 65 years-and-above age group reported the highest satisfaction levels, while the youngest age group reported the lowest satisfaction.

mules.* An account had the profile of a salaried individual earning only ₹10,000 per month, yet exhibited inflows and outflows of ₹40 lakh per day, with cash withdrawals amounting to ₹20 lakh daily. Account access was observed from five different countries, based on IP address logs, clearly indicating that it was being operated remotely. The risk in such a case is evidently not identity, but transaction patterns that are inconsistent with the customer's profile.

To detect money-laundering risk, entities therefore have little choice but to undertake transaction monitoring. It is not sufficient to rely solely on static rules; dynamic pattern analysis is essential, given the presence of sentient adversaries who adapt their behaviour based on real-world interactions. Furthermore, no financial entity—irrespective of size—can deploy large teams of human analysts to investigate alerts generated by static rules, eliminate false positives, and then prepare Suspicious Transaction Reports (STRs) for submission to the Financial Intelligence Unit.†

One of the fundamental challenges entities face is the lack of contextual information about new customers entering their systems. Without such context, it becomes difficult to assign an appropriate risk profile. The example above illustrates this problem: a newly opened account was able to launder large sums within a week without the anomaly being detected.

One possible approach to addressing this challenge is through the use of artificial intelligence. Before exploring how this might be

---

*Indian Cyber Crime Coordination Centre (I4C), *Trends in Cybercrime and Banking Abuse*, slide 13, https://ciet.ncert.gov.in/storage/app/public/files/17/Workshop_and_training_files/fsce/day1_ppt.pdf. Accessed on 25 May 2026.

†Under FATF recommendations, each country is required to establish a nodal agency to receive, process, analyse and disseminate suspicious transaction reports to intelligence and law enforcement agencies. In India, this function is performed by the Financial Intelligence Unit (FIU), which reports to the Ministry of Finance. See Financial Intelligence Unit–India, 'About FIU-IND', https://fiuindia.gov.in/files/About_FIU-IND/About_FIUIND.html for more information. Accessed on 25 May 2026.

done, it is useful to clarify what is meant by 'artificial intelligence'. Surprisingly, there is no universally agreed-upon definition of AI. The term is often used to describe a loose collection of technologies[35] that may differ significantly from one another but are grouped together for convenience.

Broadly, three types of AI are commonly discussed: generative AI, predictive AI and content-moderation AI. Generative AI is now widely recognized following the emergence of systems such as ChatGPT, which excel at producing text and images based on prompts. Predictive AI, by contrast, attempts to forecast future states to inform present decisions, but has often failed to deliver on its promises across multiple domains, despite considerable hype generated by media headlines.[36]

If one were to apply a set of techniques to analyse transaction patterns and flag potential early indicators of money laundering, could this be classified as AI? And if so, under what framework?

Arvind Narayanan and Sayash Kapoor offer a useful framework for answering this question:*

1. If a task involves work that would ordinarily require human training or expertise, and a machine can perform it, it is likely to be AI.
2. If the system's behaviour emerges from learning through examples or searching through data, it is likely to be AI.
3. If the system makes autonomous decisions and adapts to its environment, it is likely to be AI.

Analysing large volumes of raw transaction data to identify potential money-laundering patterns satisfies criteria (1) and (2) fully and (3) partially. A system built for this purpose could therefore reasonably be described as AI. The more important question, however, is how

---

*Narayanan, Arvind, and Sayash Kapoor, *AI Snake Oil: What Artificial Intelligence Can Do, What It Can't, and How to Tell the Difference*, Princeton University Press, Princeton, 2024, pp. 12–13.

such a system would be designed, and whether it would work in practice.

Financial transactions have long been modelled using graphs, where each account is represented as a node and each transaction as an edge.[37] In recent years, combining such graph structures with deep-learning models has enabled the development of techniques capable of identifying suspicious patterns.

For instance, a self-supervised approach using graph neural networks such as LaundroGraph[38] can encode customer behaviour and transaction flows within the context of AML, providing a starting point for potentially suspicious transactions by generating anomaly scores, clustering transactions and tracking customer behavioural changes over time. Similarly, techniques such as HGSuspector[39] can detect fraudulent activity within directed heterogeneous graphs. Attention-based techniques, popularized by generative AI, have also been applied to detect fraud rings and hidden networks.[40] The key requirement for such approaches is access to raw transaction data across multiple entities.

A key question then emerges: would these techniques actually work in the real world, given the repeated failures that have always accompanied predictive AI across domains?

Once again, the framework provided by Arvind Narayanan and Sayash Kapoor helps in evaluating this:*

1. Good prediction is not the same as good decision-making.
2. Opaque systems can be strategically gamed.
3. Over-reliance on AI without oversight can worsen outcomes.
4. Training data may not match the population on which the system is deployed.
5. AI systems can exacerbate inequality.

*Ibid, p. 59.

In the context of transaction analysis, AI is not making definitive predictions that a transaction is illicit, nor is it directly making enforcement decisions (such as freezing an account). It merely classifies transactions as potentially suspicious, which may then lead to the filing of a suspicious transaction report (STR) by the financial entity. Moreover, transaction modelling based on dynamic patterns—rather than static thresholds—can reduce the scope of gaming (e.g., by keeping cash transactions just below the reporting threshold).

Over-reliance on AI is not a major concern here, as it is used only to help in filing STRs; it serves as an assistive tool rather than a decision-maker. Graph neural networks, in particular, model transactions and are designed to be identity-invariant, meaning the training data would be entity- and domain-specific, and not population-specific. The more difficult question is whether their deployment may worsen inequality, especially given the widespread side-effect of debanking of vulnerable populations under static, rules-based KYC regulations. However, this impact can be mitigated through more creative regulatory approaches (discussed elsewhere in this chapter).

Optimizing Transaction Profiling Costs

Transaction profiling for every financial account comes with costs, regardless of who implements it. These fall into three categories:

1. **Hardware:** Graph neural networks need dedicated CPU and GPU resources, which typically can't be shared with core banking systems. This is because the primary purpose of a financial entity's infrastructure is to facilitate transactions reliably, which is only possible if its hardware can be profiled accurately to meet service levels. Combining hardware used for running neural networks for detecting money laundering with AI workloads is hence not workable, as it complicates workload provisioning enormously.

2. **Software:** Graph neural networks are AI models that must be continuously developed, updated and maintained, as they interact with sentient adversaries. Without regular updates, the training data quickly becomes outdated, leading to model drift and reduced effectiveness.
3. **Network Bandwidth:** One way in which these systems can be designed without interfering with core banking is to offload analysis to the public cloud and run them in batches, such as at the end of each day. While this may be computationally efficient, it introduces additional data transfer (egress) costs when moving data to and from the cloud.

Since these costs apply to every financial entity—and not all institutions have the means to absorb them—there is a strong case for pooling resources. A shared software standard and common hardware platform could help in optimizing costs. Multiple models could support this, such as a sovereign cloud operated through a subsidiary (e.g., IFTAS Cloud)[41] or an umbrella entity structure where financial entities are both users and shareholders (e.g., NPCI), funded through asset development funds or similar mechanisms.

Shared infrastructure—across software, hardware and models—is necessary but not sufficient. A key challenge is economic: accounts used solely for Direct Benefit Transfers (DBT), such as Jan Dhan Yojana accounts, do not generate revenue for financial institutions. From a purely cost perspective, running full-scale ML-based risk analysis on these accounts is therefore impractical.

The risk of money laundering in such accounts arises from several structural reasons:

1. Persistent economic inequality has left a large section of the population dependent on government handouts. They have neither assets nor jobs, nor stable incomes that meet their bare minimum needs.

2. In their drive to accelerate financial inclusion, policymakers adopted a target-driven approach to opening bank accounts, which are open-ended and can receive funds from any source.
3. While individuals are nominally restricted from opening multiple such accounts, this restriction can't effectively be enforced by the bank, unless there is a central, searchable repository of bank accounts linked to individuals. However, no such database exists since India's AML architecture remains largely top-down: institutions submit STRs to the Financial Intelligence Unit (FIU) but the FIU does not send anything back, even though it might have a system-wide view of individuals or entities across the entire financial landscape.
4. For many individuals, a bank account becomes their only economic asset—one that can be informally 'rented out' by handing over control to money launderers.

Hence, there are a few possible regulatory solutions that can work on managing these two contradictory goals (financial inclusion and mitigation of money laundering).

The first would be to remove the restriction that an individual may hold only a single Jan Dhan Yojana account, since it is unenforceable.

The second would be to introduce lifetime KYC-compliant DBT accounts. Such accounts would be restricted to receiving funds only from a whitelisted set of government sources (state and central DBT programmes), while allowing unrestricted outgoing transfers. By design, this eliminates ML risk on the inflow side completely, removing the need for account freezes or periodic KYC updates; the account could therefore remain active for the lifetime of the holder. This type of carve-out is consistent with FATF regulations and has international precedent—for example, Brazil's *Bolsa Família* programme, which distributes benefits through prepaid cards[42] to eligible families who can use them for any purpose, including cash withdrawals. In effect, restricting incoming funds to approved sources transforms the JDY account into a functional equivalent of a prepaid account, while preserving usability for beneficiaries.

The third would be to move away from coercing banks into participating in financial inclusion through target-based mandates. As discussed earlier, such an approach produces favourable short-term metrics—such as the number of accounts opened—but undermines long-term financial inclusion. Banks have little incentive to maintain these accounts, as they do not even break even for them. While targets shift the immediate cost of financial inclusion from the government to banks, they ultimately weaken institutions, reduce customer satisfaction and create distrust. Perhaps the time has come to strengthen India Post—already a licensed payment bank—as the primary vehicle for delivering direct benefit transfers, through budgetary allocation. This would make the true cost of financial inclusion transparent.

The final step would be to rethink the model of data sharing between the FIU and regulated entities. At present, the flow of information is one-way: entities submit STRs to the FIU, but receive little in return. If the objective is genuine collaboration, a two-way data-sharing framework should be explored. For instance, entities could avoid the costly process of analysing raw transactions unless prompted by a 'trip-wire' alert from the FIU based on the STR gathered from another entity. This would help detect coordinated networks of accounts created for laundering purposes and allow earlier, more targeted intervention. In such a model, the FIU would function not only as a push entity to which STRs are sent but also as a pull entity issuing alerts and asking for early warning signals to trigger deeper STR analysis.

Although FATF is the primary force that drove most of global AML regulations, India's KYC regime illustrates, not just for common citizens but for regulatory bodies and law enforcement, how frameworks fail because of the lack of systemic thinking on how to balance competing concerns by policymakers. The consequence of this lack of macro thinking is, thankfully, not irreversible. It can be fixed through a series of technical and regulatory innovations and a deliberate effort to create an ecosystem for those innovations to sustain themselves through industrial policy.

Without such deliberate crafting, KYC becomes a tool of harassment and misuse—an extortion scheme by middlemen conniving with law enforcement agencies. While it originated as part of sanctions by Western countries to impose their supremacy on others—whom they believed were their economic, military or ideological adversaries—its current implementation in India can only be categorized as insane.

Our bureaucratic apparatus has turned this into a system of self-sanctions that is arbitrary and capricious, and ended up hurting the very goals it was meant to serve—financial inclusion and formalization. Unlike other sanctions that were imposed by the Trump administration that hurt our interests and constrained our freedom of action, the KYC regime, as it exists today, is a true Made-in-India product when it comes to constraining our businesses.

The good news is that it can be fixed with very minimal costs, if the state really wants to.

2

How Industrial Policy and Social Capital Impact Sovereignty

There is one emotion that supersedes all others when an Indian citizen deals with the state—raw, unadulterated fear of harassment. It is a cultural fear, long emblematic of the extended arm of the state.

No one understands this better than cybercrime operators, who have learnt to monetize it more effectively than even the corrupt arms within the state. It often begins with a phone call. The tone is polite but firm, and the script typically unfolds as follows:

Caller: Are you Mr X, residing at this address?

X: Yes, who is this?

Caller: I am a customs officer from Mumbai port. Do you often get parcels from a 'random country'?

X: No, what is the matter?

Caller: There is a parcel containing a large quantity of narcotics, addressed to your office address and in your name. We need to investigate, and you are required to come to our office in Mumbai within the next 24 hours.

X: What? I have nothing to do with that.

Caller: Mr X, that does not matter. It has your name on it, and we must investigate. If you do not cooperate, we may have to issue an arrest warrant, and you could face life imprisonment. We just want to clear this matter once.

X (fearful): Ok, what should I do?

Caller: Please download this app (video-calling) on your phone so that we can speak immediately.

X downloads the app and connects with the caller.

Caller (on video, dressed in a customs officer uniform with an official-looking background) conducts a mock investigation and eventually suggests that the parcel may have been sent in error, and asks Mr X to submit his tax returns for the past three years.

X (relieved): I hope you now understand that I am innocent (sends tax returns).

Caller: We still need to complete the investigation, which will cost both of us time and money. If you can deposit ₹5 lakh to this UPI ID, we will issue a receipt. If the investigation concludes in your favour, the amount will be returned with 18 per cent interest.

X proceeds to make the payment.

Caller provides him an official-looking receipt with a letterhead, stamp and summary of the conversation.

But what does this incident have to do with sovereignty?

To understand that, one must first recognize that India is not self-reliant on most objects one relies on a day to day basis.

Consider solar panels or mobile phones or fertilizers or modern medicines. If one were to enumerate all the components required to make any of the above, many of these would not be available locally, but had to be imported from China.

Hence the slogan 'Made in India' is essentially absurd because in reality it is 'Assembled in India and Made in China'.

Why does this absurdity persist, even after India climbed from rank 134 (out of 189) in 2014 to 63 in 2022 in the World Bank's 'Ease of Doing Business' rankings? It turns out that not only was the methodology flawed, but also the rankings themselves were rigged to secure financial commitments from certain countries.* Put simply, the rankings did not reflect the actual ease of doing business, but rather the World Bank's ease of getting financial commitments from some countries.

There are several reasons why India is not a manufacturing powerhouse—even for something as basic as idols:†

1. **Scale:** Most Indian units operate at a small scale. Fixed costs therefore raise per-unit costs, making products less competitive.
2. **Labour Laws:** Once a factory exceeds 100 workers, compliance requirements increase significantly. Many firms lack the capital and administrative capacity to manage this.
3. **Pervasive Petty Corruption:** The British Raj may have ended, and the License Raj dismantled, but the 'Inspector Raj' persists. Rent-seeking behaviour within the bureaucracy has achieved a high art form, increasing costs and reducing competitiveness. One of the more difficult reforms—should it become politically feasible—would be in administrative law, to enable greater bureaucratic efficiency and institutional progress.
4. **Infrastructure:** Whether land, electricity or transport, infrastructure is either inadequate or unreliable. Businesses must compensate (for example, by installing diesel generators), which raises costs further.

---

*Machen, Ronald C., Matthew T. Jones, George P. Varghese, and Emily L. Stark, *Investigation of Data Irregularities in Doing Business 2018 and Doing Business 2020: Investigation Findings and Report to the Board of Executive Directors*, Wilmer Cutler Pickering Hale and Dorr LLP, 15 September 2021. Accessed on 25 May 2026.

†Contractor, Farok J., 'Thousands of Hindu Gods – Made in China: Seven Reasons Why India's Manufacturing Competitiveness Lags China's', *Rutgers Business Review*, Vol. 3, No. 1, Spring 2018, https://rbr.business.rutgers.edu/sites/default/files/documents/rbr-030101.pdf. Accessed on 25 May 2026.

Persistent petty corruption undermines manufacturing ambition unless supported by capital, connections and influence. Consider the case of Wintrack Inc., reportedly forced to shut operations in India due to constant bribery demands at Chennai customs.* Or the example of an entrepreneur attempting to manufacture drones under the Make-in-India programme, and was asked to employ an official's son as Director of Finance. But then, it is far too time-consuming and punishing for any individual or entity to take on the corrupt in Indian official systems.

In the first case, corruption was sophisticated bribery; in the second case, it was elevated into a permanent entitlement scheme within the bureaucracy. The difference between rent-seeking officials and the cybercrime operators is therefore one of scale and ambition. Bureaucrats typically demand bribes when they are approached by the citizenry; cybercriminals, however, actively reach out to the citizenry—expanding the total available market for extraction.

A similar pattern plays out in taxation as well. Consider the promise of GST, which is supposed to be a value-added tax with tax credits claimable at every stage of production. In reality, however, several cases exist where such credits cannot be fully utilized. Inverted duty structures often prevent full credit claims (e.g., when goods taxed at 5 per cent can't offset tax paid on higher-taxed input services). This distorts cost structures and increases the cost of manufacturing in India.

The correct way to look at Indian manufacturing, therefore, is not through the perceived 'ease of doing business', but through a realistic 'unease of doing business.'† Harassed by everyone and everything, the

---

*Hinduja, Udit, 'Wintrack Founder's Spat with Chennai Customs Started in January. He Had Ordered Handcuffs', *The Print*, 3 October 2025, https://theprint.in/feature/wintrack-founder-chennai-customs-january-spat/2756575. Accessed on 25 May 2026.

†Mukherjee, Andy, 'The Great Unease of Doing Business in India', *Bloomberg*, 17 January 2025, https://www.bloomberg.com/opinion/articles/2025-01-16/great-unease-of-doing-business-in-india. Accessed on 25 May 2026.

Indian entrepreneur is resilient and somehow survives in most cases but often has little capacity left to scale or innovate.

This is why atma-nirbhar is not a reality even for most objects or components made in India.

The decisions and behaviour of the Indian State and its bureaucracy constitute what we call industrial policy, or simply 'policy'. China's emergence as a global powerhouse—capable of producing everything from nuts, bolts and screws to ships at low cost—reflects deliberate state action to support and scale domestic industry for global supply. In contrast, policies in India have often discouraged local businesses from becoming reliable and cost-effective suppliers, even for assured domestic demand, such as the production of idols.

The weakest link—corruption, which is often ignored—drains societal energy. Even sophisticated forms of corruption continue to evolve, adapting to modern structures.

Policy success or failure thus is an intrinsic part of why some states succeed at scale and why others fail. This has been true for centuries, going back to the rise of the British Empire.

The Early Industrial Empire States

There was a time when power meant land. For centuries, wealth came primarily from agriculture. The richest nations were those with the most fertile land, strongest armies to defend it and the most efficient systems to extract labour from those who worked on it.

Then, in late 18th-century Britain,[43] a new development rewrote the rules of power forever. It began with machines—simple, unassuming contraptions designed to replace human effort. The spinning jenny accelerated textile production. The steam engine converted coal into motion, enabling railways. Soon, mills and factories emerged along rivers, their chimneys darkening the English sky. The world had entered the Age of Industry.

For Britain, this was not just economic progress—it was the foundation of an empire built not merely on land, but on manufacturing. Textiles produced in Manchester were shipped to every corner of the world, while steel from Sheffield laid railway tracks in colonies thousands of miles away. The British Empire was no longer defined solely by conquest; it was driven by industry—by the ability to transform raw materials into wealth on an unprecedented scale.

Europe watched and followed. France, Belgium and Germany developed their own factories, expanded their railways and pursued their own industrial ambitions. Power was no longer held exclusively by kings and emperors alone, but increasingly by merchants and industrialists. It shifted towards those who controlled machines, coal and trade routes.

After the American War of Independence, the United States—initially agrarian—recognized the need to become an industrial power. Alexander Hamilton[44] set up the Society for Establishing Useful Manufactures in 1791 to achieve independence from British imports. By the late 19th century, the US had become an industrial giant. Where Britain had built its power on mechanization, America took it further—it mastered scale. Steel mills in Pittsburgh, oil refineries in Texas, and automobile plants in Detroit facilitated a new kind of industrial dominance.

Henry Ford introduced the assembly line, making it possible to build cars faster and more cheaply than anyone had imagined. The railroad network stitched together the vast continent, allowing goods and raw materials to move at unprecedented speeds. The US was not just manufacturing—it was mass-producing the future.

By the end of the Second World War, Europe lay devastated and relied on American support for reconstruction. Through the Marshal Plan,[45] the US provided grants and loans amounting to $29.3 billion (in 2005 values) to rebuild its industries. Japan, though also shattered by war, with its cities in ruins and its economy in the craters, retained much of its industrial base, unlike Europe. Initially, the US occupation imposed sweeping military, economic and social reforms. However, the emergence of communism—culminating in the Cold War—and the Korean War promoted a strategic shift, leading the US to channel approximately $15.2 billion (in 2005 value)* into rebuilding Japan's industrial capacity.

Japan recognized that the future of industry was not just about producing more, but about producing better. While the US focused on mass production, Japan emphasized precision. In the 1950s and 60s, Japanese firms developed manufacturing techniques that prioritized efficiency and quality. Toyota revolutionized the automobile industry through lean manufacturing, reducing waste and improving reliability, while companies such as Sony and Panasonic perfected consumer electronics, establishing Japan as a global leader in televisions, radios and cameras.

By the 1980s, Japanese cars were no longer seen as cheaper alternatives to American models—they were often considered superior. Factories that had once been dismissed as mere imitators became the gold standard of manufacturing. South Korea and Taiwan soon followed. Within a few decades, they built industries that rivalled those of the West: Samsung and Hyundai became household names globally, while Taiwan's semiconductor industry quietly became the backbone of modern electronics.

Rise of the Chinese Industrial State

In the 18th century, China under the Qing dynasty was among the most technologically advanced societies in the world. It had the largest iron and steel production during the Song dynasty (its

*Serafino, Nina, Curt Tarnoff, and Dick K. Nanto, *U.S. Occupation Assistance: Iraq, Germany and Japan Compared*, CRS Report for Congress, 23 March 2006, p. 11, https://www.history.navy.mil/research/library/online-reading-room/title-list-alphabetically/u/us-occupation-assistance-iraq-germany-japan-compared.html#page11. Accessed on 25 May 2026.

output rivalled that of 18th-century Britain)* and had pioneered paper money, printing and gunpowder, developed sophisticated engineering capabilities and sustained a vast trading network—earlier exemplified by Admiral Zheng. He voyaged to Africa in the early 15th century. Yet from the late 15th century onwards, China turned inward. The Ming and later Qing dynasties restricted overseas trade, curtailed large-scale shipbuilding, and slowed technological advancement. By the 18th century, Europe had begun to industrialize, while China remained largely agrarian.

The Opium Wars (1839–42; 1856–60) marked a decisive rupture. British forces, powered by steam and industrial technology, defeated China, forcing it to open its markets, cede Hong Kong and accept foreign economic control. China shifted from being a manufacturing power to a semi-colonial economy supplying raw materials to industrialized nations.

In the late 19th century, the Qing state attempted reform through the Self-Strengthening Movement (1861–95), establishing modern shipyards, steel factories, telegraph lines, railways and Western-style military institutions. However, these efforts remained partial and inconsistent. The ruling elite feared that industrialization might weaken central authority by empowering regional actors, while corruption and bureaucratic inertia stalled progress. By the early 20th century, Japan—once less advanced—had modernized rapidly and defeated China in the First Sino-Japanese War (1894–95). The fall of the Qing dynasty in 1911, followed by decades of warlordism, civil war and Japanese invasion (1937–45), further disrupted industrial development. By 1949, when Mao Zedong and the Communist Party came to power, China's industrial output accounted for less than 1 per cent of global manufacturing, and the country remained predominantly agrarian.

---

*McDermott, Joseph P., and Yoshinobu Shiba, 'Economic Change in China, 960–1279', in John W. Chaffee and Denis Twitchett (eds), *The Cambridge History of China*, Vol. 5, Part 2: *Sung China, 960–1279*, Cambridge University Press, Cambridge, 2015.

Mao's government pursued state-led industrialization. The First Five-Year Plan (1953–57) focused on heavy industry—steel, coal and machinery—through state-owned enterprises and Soviet-style central planning. While it expanded basic industrial capacity, the system remained inefficient. The subsequent Great Leap Forward (1958–61) sought to accelerate industrialization but proved catastrophic: misguided policies, including backyard steel production and disruptive agricultural reforms, led to economic collapse and widespread famine. By the 1970s, China had built a large but inefficient industrial base, governed more by bureaucratic targets than market signals, even as Japan and South Korea advanced through export-oriented, market-driven strategies.

- A decisive shift came under Deng Xiaoping, who introduced market-oriented reforms and opened China to global trade and investment. Special Economic Zones (SEZs) like Shenzhen attracted foreign manufacturing.
- Private enterprises were encouraged under state oversight.
- Foreign companies were allowed in but had to partner with domestic firms (ensuring tech transfer).
- Labour-intensive manufacturing expanded rapidly, making China the world's low-cost factory.

During the 1990s, China's GDP grew at an average rate of 9–10 per cent annually.* It became a leading manufacturer of electronics, textiles and consumer goods. By the early 2000s, China had evolved beyond low-cost production into a global industrial hub, supported by policies such as:

- Made-in-China Strategy (2006)—a strategic shift from low-end manufacturing towards higher-value industries.
- Infrastructure Growth—large-scale investment in roads, ports and high-speed rail.

---

*World Bank Group, 'China', https://www.worldbank.org/en/country/china. Accessed on 25 May 2026.

- Strategic Acquisition of Western Firms—buying foreign companies to gain technological know-how.

This gave rise to globally competitive companies such as Huawei, Lenovo and BYD, while China emerged as the largest producer of steel, solar panels and electronics.

The 'Made in China 2025'[46] plan aims to make China self-sufficient in semiconductors (reducing dependence on US chips), AI and automation (competing with Silicon Valley), and electric vehicles and green energy (dominating future industries).

The Chinese Envy and Its Strange Side Effects

The rise of China as an industrial and technological power evokes strong reactions across the world. To some, it represents a decisive force reshaping global power dynamics—a nation that has mastered industrial dominance. To others, it is a cautionary example of authoritarian capitalism, where economic strength has come at the cost of individual freedoms. Yet, beyond these ideological positions lies a clear reality: China has emerged as a formidable power across both traditional and emerging sectors.

While the instinctive rejection of its successes as illegitimate, propped up by state subsidies and unfair practices, persists in some quarters, others—India included—have often attempted to replicate its trajectory by copying policies, assuming that what worked for Beijing will work equally well for New Delhi.

Over the past four decades, China has pursued a long game, methodically shaping its economic structure with a singular focus on enhancing national power, much like earlier industrial states. It built infrastructure at an unprecedented scale, invested heavily in core industrial capacity even as others shifted towards services, and implemented policies that ensured not just competitiveness but total dominance in critical sectors. Rather than waiting for global markets to become fair, it actively shaped them to its advantage.

Today, China is indispensable to global manufacturing supply chains, accounting for over 30 per cent of world production. It leads in renewable energy[47] and dominates sectors such as 5G, rare earth processing[48] and electric vehicles. This position is not accidental, but the outcome of a disciplined, long-term industrial strategy.

Its rise was also facilitated by the increasing 'financialization' of the United States and other advanced economies, which offshored manufacturing to China in the belief that their comparative advantage lay in design, branding and innovation. Over time, this eroded their industrial capabilities. As China consolidated control over supply chains, many countries that once prided themselves on manufacturing excellence found themselves dependent on it.

India, by contrast, liberalized its economy in 1991 but did not pursue an industrial strategy with comparable clarity or consistency. While liberalization marked a turning point, it remained incomplete. India embraced services while underinvesting in manufacturing, excelled in software while neglecting hardware, and allowed private enterprise to grow without embedding it within a deeper industrial framework. As a result, despite its strength in IT services, India remains dependent on global supply chains for critical technologies, from semiconductors[49] to advanced industrial machinery. In a world where manufacturing capacity, supply chain resilience and computing power shape geopolitical influence, this dependency poses a structural disadvantage.

It is easy to argue that India cannot match China's scale of investment or its ability to design and execute industrial policy with similar speed and cohesion. The Chinese state has the capacity to inject billions into strategic sectors, direct capital without the constraints of democratic consensus, and implement projects at a pace that is often unthinkable within India's political and bureaucratic framework. Yet financial scale alone is not the only metric of success.

If it cannot centralize control, it must leverage its strengths—its demographic dividend, entrepreneurial private sector, expanding consumer market and its capacity to build global goodwill in a way that China never could.

For this, industrial policy must evolve beyond a collection of schemes and incentives into a coherent doctrine—a structured vision that aligns national resources with long-term national ambition. It is not about slogans or episodic enthusiasm for emerging sectors, but about creating the conditions for sustained global leadership by Indian enterprises.

This is where urgency is lacking. India's past successes—the Green Revolution, the White Revolution, the establishment of C-DAC and self-reliance in pharmaceuticals—were born out of necessity and driven by catalytic urgency. The state did not wait for markets to correct themselves; it intervened with clarity and purpose.

Today, despite significant movement in policy, that sense of existential urgency is largely absent. Initiatives such as the semiconductor push, Production-Linked Incentive schemes, and efforts towards indigenous AI development are steps in the right direction, but they do not yet reflect a coherent doctrine or structured vision. Without such catalytic force, India risks meandering—making incremental progress while the rest of the world advances in quantum leaps.

No industrial policy, however well-intentioned, can succeed in isolation. Without a complete overhaul of India's education system and STEM ecosystem, the ambitions of technological advancement will remain limited. Nations that lead industrial revolutions are not defined by capital alone, but by the strength of their talent pipelines designed for cutting-edge innovation. China, despite its so-called censorship-driven authoritarian structure, has built world-class research institutions, consistently producing high-performing STEM graduates and generating patents and deep-tech innovation at scale. In contrast, India's engineering education often remains outdated, producing graduates in large numbers but with limited exposure to frontier technologies.

If India does not invest in fundamental education reform, it will remain a consumer, not a producer, of future technologies. Its startups will continue to depend on global capital, its industries will be reliant on foreign intellectual property, and its policymakers will remain reactive rather than proactive. A nation that does not control its technological future cannot be truly sovereign. The race for industrial supremacy is not just about infrastructure, capital or policy; it is about human capital—the ability to innovate at the frontiers of science and technology.

This is not an argument for mimicking China, nor an attempt to paint an alarmist picture of India's position in the global order. In geopolitics and global economics, sentiment is irrelevant—outcomes are what matter. The intent here is analytical: to seek strategic clarity by examining how global industrial powers have historically built dominance through coordinated reforms across institutions, sustained investments and the alignment of domestic capabilities with foreign and trade policies.

What Is Industry Policy?

If industry policy is neither a collection of schemes nor a set of incentives, what is it then? A clear definition is provided by Barry Naughton:* 'Industrial policy is any type of selective, targeted government intervention that attempts to alter the sectoral structure of production toward sectors that are expected to offer better growth than would occur in the (non-interventionist) market equilibrium.' He further distinguishes policy from mere pronouncements or target-setting; it must be backed by a significant commitment of resources. The emphasis here is on 'commitment' and not just 'allocation', given the familiar tendency within Indian bureaucracy to allocate budgets without fully deploying them—rendering the policy dead on arrival.

*Naughton, Barry, *The Rise of China's Industrial Policy, 1978 to 2020*, Academic Network of Latin America, 2021, p. 18.

Such commitment, however, implies that resources are available as 'surplus' for policymakers in the first place. It would also be important to recognize that the resources required for policy success extend beyond capital, land or incentives; they include an existing industrial base that can support expansion into new areas. Without such a foundation, attempts to alter the structure of production will always fail because there is no underlying capacity on which to build.

For instance, consider the tariff measures introduced by President Trump in his second term. As the administration's Treasury Secretary, Scott Bessent, said:* 'On one side, the president is reordering trade. On the other side, we are shedding excess labor in the federal government and bringing down federal borrowings. That will give us the labor that we need for the new manufacturing'. This view was widely mocked, including through a popular AI-generated video† depicting American workers ruefully making inner garments in a textile manufacturing shop. The critique resonated because it highlighted a fundamental point: industries cannot be built through disruptive measures alone (e.g., firing federal workers, imposing tariffs), hoping that a sector will somehow restructure itself to manufacture a good (clothing) without the foundational infrastructure and ecosystem required to sustain production.

Reshoring or reindustrialization, even with substantial state support, is not easily achievable—as illustrated by Foxconn's proposed LCD plant in Wisconsin.‡ The company was promised US$3 billion in subsidies for creating at least 13,000 jobs, against an

---

*Lane, Sylvan, 'Bessent: Federal Layoffs Will Help Fill Factory Jobs Created by Trump Tariffs', *The Hill*, 4 July 2025, https://thehill.com/business/5236250-trump-tariffs-economic-recovery/. Accessed on 25 May 2026.

†Wilkins, Joe, 'China Creates Mocking AI Video of Average Americans Working in Garment Factory', *Yahoo News*, 8 April 2025, https://www.yahoo.com/news/china-creates-mocking-ai-video-182144462.html. Accessed on 25 May 2026.

‡Bauerlein, Valerie, 'Foxconn Says It Will Move Forward With Wisconsin Plant After Conversation with Trump', *The Wall Street Journal*, 1 February 2019, https://www.wsj.com/articles/foxconn-says-it-will-move-forward-with-wisconsin-plant-after-conversation-with-trump-11549047068. Accessed on 25 May 2026.

investment commitment of US$10 billion. It then scaled back these commitments, citing high labour costs and limited commercial viability,* and instead offered to build a vaguely defined 'AI, 8K + 5G' knowledge park.†

The decline of Intel from its pole position offers further insight into why industrial capabilities once lost are almost impossible to get back. Intel's claim to fame was its x86 microprocessors, based on CISC (Complex Instruction Set Computer) architecture. While less streamlined than the RISC (Reduced Instruction Set Computer) architecture used by competitors, it was very cheap to manufacture at scale, allowing Intel to dominate the mass market. Profits were reinvested in research and development of its manufacturing technology, enabling the company to pack more transistors and thus deliver higher processing power, often maintaining or exceeding competitors' prices while outperforming them. This approach allowed Intel to become a near-monopoly in not just mass-market desktop chips but also in server chips, the latter offering considerably higher margins.

These margins were driven not only by monopoly but also because of manufacturing yield—the proportion of wafers successfully converted into usable chips. Intel owned its foundries and specialized in vertical integration; given that a foundry producing chips requires several billion dollars of investment, not only for setup but also for constant upgrades on process technology, higher volumes reduce per-unit costs and improve profitability, as the cost to make a chip decreases. Chip manufacturing thus scales effectively under two conditions: high production volumes that

---

*Yu, Jess Macy, and Karl Plume, 'Exclusive: Foxconn Reconsidering Plans to Make LCD Panels at Wisconsin Plant', *Reuters*, 31 January 2019, https://www.reuters.com/article/technology/exclusive-foxconn-reconsidering-plans-to-make-lcd-panels-at-wisconsin-plant-idUSKCN1PO357/. Accessed on 25 May 2026.

†'Everything You Need to Know about Foxconn in Wisconsin', *The Verge*, 18 September 2025, https://www.theverge.com/policy/2019/4/26/18518650/foxconn-wisconsin-factory-lcd-buildings-us. Accessed on 25 May 2026.

improve efficiency over time, and sustained reinvestment of profits into next-generation process technologies.

The tight integration between manufacturing and design allowed yields to improve through standardization. For instance, many chip layouts followed grid-based, uniform designs, which imposed constraints on chip geometries. While these constraints did not significantly affect x86 CPUs, they became limiting when Intel attempted to diversify into other categories, such as GPUs or mobile chips. Relaxing these constraints improved performance but reduced yield and increased costs. These trade-offs contributed to a shift towards financialization: prioritizing high-margin x86 server and desktop chips, while forgoing opportunities in emerging segments with lower margins, alongside an increasing focus on stock buybacks and dividends.

This mindset also led Intel to pass on manufacturing CPUs for the iPhone. The company concluded that the offered price was below its projected costs and that volumes would be insufficient. As former CEO Paul Otellini noted,* 'At the end of the day, there was a chip that they were interested in that they wanted to pay a certain price for and not a nickel more, and that price was below our forecasted cost. I couldn't see it. It wasn't one of these things you can make up on volume. And in hindsight, the forecasted cost was wrong, and the volume was 100x what anyone thought'.

Apple initially turned to Samsung for ARM-based System-on-Chip (SoC) designs, and later, as Samsung became a competitor, moved to in-house chip design with fabrication by TSMC. As the business grew manifold over time, TSMC advanced its process technology faster than Intel's in-house foundries. When Intel eventually considered outsourcing fabrication to TSMC, the Biden

*Madrigal, Alexis C., 'Paul Otellini's Intel: Can the Company That Built the Future Survive It?', *The Atlantic*, 16 May 2013, https://www.theatlantic.com/technology/archive/2013/05/paul-otellinis-intel-can-the-company-that-built-the-future-survive-it/275825/. Accessed on 25 May 2026.

administration intervened with US$7.86 billion* under the CHIPS Act to revive domestic semiconductor manufacturing. Yet even this proved insufficient: leadership changes followed, and Intel has since contemplated scaling down or exiting parts of its manufacturing operations.†

The Intel case underscores a broader point: industrial policy succeeds only when there is an existing base of firms producing what markets demand, generating surplus and reinvesting in R&D. Where firms prioritize margin preservation—through buybacks, dividends and dominance in legacy segments—they risk falling behind in emerging technologies. Monopoly power, paradoxically, can create internal disincentives to pursue new opportunities, even for companies built on a culture of constant vigilance.

The Chinese Industry Policy

Barry Naughton‡ makes a compelling case that China never had a formal industry policy until around 2006, with more explicit articulation emerging after 2010—outcomes that have not been adequately studied and are still unknown. If it never had an industry policy till 2006, how did China become an industrial powerhouse? To explain this, Naughton points to the successive waves of market-oriented reforms that energized the economy.§

The first wave focused on agriculture. Constraints on farmers were removed, allowing them full freedom over how to use their land. Procurement quotas and price controls were slowly

---

*Intel Corporation, 'Intel, Biden-Harris Administration Finalize $7.86 Billion Funding Award Under US CHIPS Act', *Intel Newsroom*, 26 November 2024, https://newsroom.intel.com/corporate/intel-chips-act. Accessed on 25 May 2026.

†Acton, Michael, 'What Pat Gelsinger's Exit Means for Intel and the US Chips Act', *Financial Times*, December 2024, https://www.ft.com/content/b9f847c6-eeb2-40f1-96fa-6ac6e647a7bf. Accessed on 25 May 2026.

‡Naughton, Barry, *The Rise of China's Industrial Policy, 1978 to 2020*, Academic Network of Latin America, 2021, p. 20.

§Ibid, p. 38.

dismantled,* and by 1993, 65 to 75 per cent of urban residents were purchasing grains from open markets rather than state stores. This led to record harvests and created a significant household surplus in the rural economy.

The second wave was built on this surplus. Rising rural incomes created demand for consumer goods, which in turn funded investment in rural enterprises. At the same time, surplus labour shifted from agriculture to non-agricultural tasks; these rural enterprises began to erode the dominance of the state-owned enterprises (SOEs), and within a decade accounted for around 30 per cent of industrial output and 25 per cent of exports.†

The third wave extended reforms to the urban economy, easing controls and enabling a rapid expansion of SMEs. This led to a surge in both the services industry and small-scale industrial units. The combined effect of the second and third waves created severe pressure on SOEs, to the point where their aggregate profits fell to near zero.‡

As policymakers simultaneously worked to create a financial sector that could sustain reforms such as banking, taxation and stock exchange institutions, several of these SMEs and rural enterprises were able to find capital and expand their business operations. This allowed them to absorb some of the labour displaced from SOEs that were not able to survive in the intensely competitive market landscape. Although this fourth wave caused severe distress for labourers who could not be reabsorbed and exited the formal labour market, surviving SOEs were restructured into joint-stock companies and eventually became profitable again.

---

*Wu, Harry X., *Reform in China's Agriculture – Trade Implications*, Briefing Paper Series No. 9, Department of Foreign Affairs and Trade, Australia, December 1997, p. 15, https://www.dfat.gov.au/sites/default/files/eaaubp9.pdf. Accessed on 25 May 2026.

†Ibid, p. 17.

‡Naughton, Barry, *The Rise of China's Industrial Policy, 1978 to 2020*, Academic Network of Latin America, 2021, pp. 38–42. (Describes the seven waves of enlivening in much more detail.)

The fifth wave was marked by large-scale urban migration, which not only accelerated growth in urban areas but also in the rural areas where migrants were able to funnel their surplus savings through transfers. The sixth wave pushed housing reform: SOE-owned housing units were privatized and transferred to workers and staff living in them. As these assets began to appreciate and could be traded or upgraded, it triggered a housing boom.

The final wave followed China's accession to the World Trade Organization in 2001, which opened global markets to Chinese exports and cemented China's position as an export powerhouse.*

In essence, China's rise was not initially driven by formal industrial policy, but by market-oriented reforms that removed existing structural constraints, allowing the Chinese economy to bloom. As Lu-Feng points out,† China's industrial capacity was not built solely through technological transfer (though it did play a part), but through reforms that were deeply absorbed by society. At its core, this capacity is organizational and societal, built over decades and shaped by the sacrifices of at least two generations. Hence, this base is not easily replicable in countries that lack similar societal and organizational capacity, irrespective of tariffs (on Chinese exports) or subsidies (given by other countries).

The cases of Intel and Foxconn's failed LCD plant in Wisconsin lend credence to Lu-Feng's hypothesis that industrial capacity is an emergent phenomenon arising from the intersection of multiple factors such as investment in education, infrastructure and sustained—often painful—market reforms, all underpinned by deep organizational and societal capacity.

---

*Ibid.

†Chan, Kyle, Thomas des Garets Geddes, and Ailsa Brown, 'Chinese Industrial Maximalism: Lu Feng', *High Capacity*, 19 June 2025, https://www.highcapacity.org/p/chinese-industrial-maximalism. Accessed on 25 May 2026.

US Export Control and Emergence of DeepSeek

Tariffs are taxation measures that nation-states have historically used as a targeted tool to protect domestic industries from external players who have either a superior product or a lower cost structure for a similar product. Tariffs are not supposed to be forever. Sanctions, by contrast, are a blunt instrument used to restrict or prohibit an entity, individual or group from engaging in business with the imposing country, often for an indefinite period. Export controls specifically restrict the transfer of certain goods or services, with the explicit intent of denying access.

It is possible to combine all these to craft a coordinated policy of denial aimed at weakening an adversary's industrial capabilities. The United States pursued such a strategy through successive measures introduced in October 2022, October 2023 and December 2024 under the Biden administration, followed by additional restrictions in March 2025 (by the Trump administration). These actions had well-defined objectives:*

- Shut down China's access to advanced AI Chips. (Denial)
- Prevent the development of local alternatives. (Deter)
- Create a dual class of normal and advanced chips and allow sales of normal chips to reduce impacts on US companies that sell these chips to China. (Self-loss mitigation)

While these measures had an impact initially, they also triggered a range of (non-exhaustive) countermeasures by the Chinese government to mitigate the impact:†

*Allen, Gregory C., 'Understanding the Biden Administration's Updated Export Controls,' *Center for Strategic and International Studies (CSIS)*, 11 December 2024, https://www.csis.org/analysis/understanding-biden-administrations-updated-export-controls. Accessed on 25 May 2026.

†Shivakumar, Sujai, Charles Wessner, and Thomas Howell, *Balancing the Ledger: Export Controls on U.S. Chip Technology to China*, Center for Strategic and International Studies (CSIS), February 2024, https://csis-website-prod.s3.amazonaws.com/s3fs-public/2024-02/240221_Shivakumar_Balancing_Ledger.pdf. Accessed on 25 May 2026.

- A state-sponsored effort supported by generous subsidies, loans and equity arrangements, to build capabilities across the advanced chip semiconductor ecosystem, including research, development, design and fabrication.
- Workarounds to export controls through networks of third parties, shell companies in permissive jurisdictions, which then buy and route the restricted items via re-packaging them with different names, and stockpiling strategies that exploit gaps in these restrictions.
- Aggressive talent acquisition by reactivating the Thousand Talents Program.*

That some of these countermeasures worked became evident two years later, when Huawei was able to launch its Mate 60 Pro smartphone† powered by advanced logic chips—likely produced by SMIC, a local company. Meanwhile, ChangXin Memory Technologies grew from zero to around 5 per cent of the global memory chip share within five years, while the Alibaba group unveiled the C930, a high-performance RISC processor for advanced applications. Yet none of these achievements came close to the breakthrough of DeepSeek, an open-source LLM developed in China using older NVIDIA Chips at a fraction of the cost incurred by other players like OpenAI, Meta, Microsoft or Google.

Le-Fung‡ points out that DeepSeek must not be viewed

*The Thousand Talents Program is an initiative designed to attract overseas expertise through a visa framework comparable to schemes in other countries (such as Australia's Exceptional Talent visa and Dubai's Golden Visa). Its objective is to encourage the migration of high-skilled talent in critical sectors.

†Pan, Che, 'While under US Sanctions, Where Did Huawei Get the Advanced Chips for Its Latest Mate 60 Pro Smartphone?', *South China Morning Post*, 2 September 2023, https://www.scmp.com/tech/big-tech/article/3233166/while-under-us-sanctions-where-did-huawei-get-advanced-chips-its-latest-mate-60-pro-smartphone. Accessed on 25 May 2026.

‡Feng, Lu, 'Chinese Industrial Maximalism', *Guancha*, 29 May 2025, https://www.guancha.cn/lufeng2/2025_05_29_777566_s.shtml. Accessed on 25 May 2026. (Translated to English via Google Translate.)

as an outlier, but as a natural outcome of a large-scale industry ecosystem spanning multiple sectors where China holds leadership positions. The sanctions regime imposed by the US and its allies was based on the flawed assumption that cutting-edge, proprietary models—which require high-cost, powerful chips—are essential for developing frontier AI systems. DeepSeek shattered this assumption by demonstrating that capable models could be built using less powerful hardware and an open-source approach. In this sense, the competition between China and the US in semiconductors and AI is not just technological; it reflects a deeper contest between different economic models of development, referred to as 'Industrial Socialism' versus 'Financial Capitalism'.

The efficient market theory postulates that capital is allocated to its most productive uses because prices reflect all available information. In such a framework, industrial growth (or de-growth) is influenced by capital flows in a market economy based on the market's expected return on investment and the perceived risks. Intel's struggle to rebuild its process-technology leadership illustrates the constraints of this model. As a publicly listed company, it can't sustain prolonged losses without pressure from shareholders demanding better returns—pressure that contributed to leadership changes* (CEO Pat Gelsinger was forced out) when its foundry strategy failed to deliver results. Left to itself, the market incentivizes firms to optimize for short-term high returns (as Intel did), even at the cost of hollowing out long-term industrial capacity.

The decline of Boeing further illustrates the risks of an excessive focus on short-term shareholder value. The company that built the 747 was once defined by an engineering-led culture

---

*Cherney, Max A., Stephen Nellis, and Anirban Sen, 'Intel CEO Gelsinger Forced Out after Board Lost Confidence in Turnaround Plan', *Reuters*, 3 December 2024, https://www.reuters.com/business/intel-ceo-pat-gelsinger-retire-2024-12-02/. Accessed on 25 May 2026.

that prioritized reliability, creating one of the most iconic aircraft in history. This, in turn, generated long-term value—not only for shareholders but also for an entire ecosystem of suppliers supporting the hundreds of thousands of components required in aircraft manufacturing.

However, following its merger with McDonnell Douglas, the combined entity saw a shift in leadership priorities. Executives increasingly focused on profit extraction rather than product excellence, emphasizing share buybacks and maximizing returns from existing models* instead of investing in new ones.

As financialization took hold, Boeing began to view itself less as a manufacturing firm and more as a sales and marketing entity overseeing an assembly operation dependent on a globally dispersed supply chain. Over time, production capabilities were progressively outsourced, hollowing out the firm's internal manufacturing base.† Notably, these risks had been anticipated in an internal white paper by engineer Alan Mulally and others (often associated with Hart-Smith's‡ critiques), yet the trajectory remained unchanged—highlighting how financialization can follow industrial success when left unchecked by market forces.

A similar pattern was evident at General Electric under Jack Welch. The company shifted from manufacturing to financial services, with engineering taking a back seat as GE Capital grew into a dominant profit centre. By the time of the global financial

---

*George, Bill, 'Why Boeing's Problems with the 737 MAX Began More Than 25 Years Ago', *Working Knowledge*, Harvard Business School, 24 January 2024, https://www.library.hbs.edu/working-knowledge/why-boeings-problems-with-737-max-began-more-than-25-years-ago. Accessed on 25 May 2026.

†Useem, Jerry, 'Boeing and the Dark Age of American Manufacturing', *The Atlantic*, 20 April 2024, https://www.theatlantic.com/ideas/archive/2024/04/boeing-corporate-america-manufacturing/678137/. Accessed on 25 May 2026.

‡Hart-Smith, L. J., *Out-Sourced Profits – The Cornerstone of Successful Subcontracting*, Boeing Paper MDC 00K0059, presented at the Boeing Third Annual Technical Excellence Symposium (TATE), St. Louis, Missouri, 14–15 February 2001, https://www.are.na/block/9268474. Accessed on 25 May 2026.

crisis, financial services accounted for nearly half of GE's profits, effectively turning it into a 'shadow bank' and had to be rescued by the FDIC.* The crisis exposed the fragility of this model, forcing intervention and ultimately leading to the divestment of its financial services arm.†

The Chinese solution to the oft-treaded path of financialization following industrialization in a market economy has been what may be described as 'Industrial Socialism'. While private economy and enterprises are encouraged, the state retains decisive control to ensure that capital does not independently dictate or direct the path of economic development.

A widely cited example of this approach is the halting of Ant Group's‡ IPO by Chinese Premier Xi-Jinping. After Jack Ma publicly criticized financial regulators in a high-profile business conference in Shanghai by saying, 'Today's financial system must be reformed. Right now, our capacity for "control" grows stronger, while our capacity to "monitor" is obviously lacking. Innovation doesn't fear regulation, but it does fear regulation by yesterday's methods,' the Chinese leadership intervened to suspend the listing. This was seen as a direct attempt by a high-profile and successful businessman—with a massive following—to dictate policy in China because of his business success, which was fundamentally at odds with the hierarchical order underpinning its policy of Industrial Socialism.

---

*'The Pitfalls of the "Financialization" of American Business', *Knowledge at Wharton*, 28 June 2016, https://knowledge.wharton.upenn.edu/article/pitfalls-financialization-american-business/. Accessed on 25 May 2026.

†Surowiecki, James, 'Back to Basics', *The New Yorker*, 27 April 2015, https://www.newyorker.com/magazine/2015/05/04/back-to-basics-why-g-e-ditched-finance. Accessed on 25 May 2026.

‡McMorrow, Ryan, and Sun Yu, 'The Vanishing Billionaire: How Jack Ma Fell Foul of Xi Jinping', *Financial Times*, 15 April 2021, https://www.ft.com/content/1fe0559f-de6d-490e-b312-abba0181da1f. Accessed on 25 May 2026.

Why (Over) Financialization Wrecks Industrialization

There is a foundational contradiction at the heart of modern capitalism: the more it talks about 'unlocking value', the less it builds anything of lasting worth. Across boardrooms, investor presentations, retail expectations, conference panels and policy documents, the vocabulary of finance has come to overshadow the grammar of production. Yet history reminds us—often starkly—that the health of a nation's real economy rests not on the churn of capital markets, but on the slow, unglamorous work of building industrial depth.

One is reminded of Britain's railways in the late 19th century—a time when financiers, rather than engineers, began to shape an industry that had once transformed the nation. Railway companies—initially celebrated for connecting distant regions and powering the Industrial Revolution—increasingly came to serve the logic of the stock exchange over that of steam and steel. A similar pattern emerged in post-war America, when the automobile industry—once the pinnacle of manufacturing ingenuity—fell under the sway of quarterly earnings. Instead of investing in fuel efficiency or safety, Detroit's Big Three prioritized share buybacks and superficial design changes to satisfy Wall Street. Meanwhile, competitors in Europe and Japan invested steadily in engineering innovation. The result was predictable: American cars became glossier but less reliable, while foreign automakers set new standards for quality. By the 1970s, factories that once symbolized industrial strength stood diminished, casualties of a system that prioritized share prices over production. These episodes illustrate a recurring truth: value, when pursued in purely financial terms, rarely translates into durable assets, resilient industries or shared prosperity.

At its core, financialization reshapes the incentives that guide entrepreneurial and policy choices. It privileges liquidity over durability, and quarterly returns over patient capital. When executives are rewarded more for stock performance than for building productive capacity, investment horizons contract. When banks find it easier to finance consumption than capital formation, the real economy is deprived of credit. And when policymakers treat rising stock indices as evidence of national strength, they risk mistaking sentiment for substance.

Contrast this with the post-war industrial strategies that rebuilt Europe and Japan, or the state-guided upgrading that transformed South Korea, Taiwan and, later, China. In each case, financial markets served industrial policy, not the other way around. Credit was channelled into targeted sectors, with export discipline and technological upgrading as explicit goals. Speculation was restrained, while manufacturing depth and innovation were actively encouraged. The outcome was not just higher GDP, but the rise of national champions whose products shaped global markets.

The industrial ascent of post-war West Germany offers a striking example. In the ruins of the 1940s, the German state rebuilt its *Mittelstand*—thousands of specialized, export-oriented manufacturers. These firms relied not on speculative capital, but on patient local banks committed to long-term industrial health. Their success was measured not in quarterly margins, but in sustained global competitiveness.

South Korea's *chaebol*s followed a different but equally deliberate path. The state directed credit, imposed export discipline, and shielded key industries from short-term market pressures. The result was the emergence of firms capable of building ships, manufacturing semiconductors, and, today, leading in advanced sectors such as electric batteries.

The United States itself offers an earlier example through the New Deal and wartime mobilization. State-led industrial policy—not market sentiment—drove the creation of infrastructure such as the Hoover Dam and complex systems like the B-29 bomber. Finance acted as an enabler, not the driver. These investments generated employment, technological spillovers and supply chains whose benefits extended well beyond their immediate economic returns.

India, too, has experienced moments of industrial ambition. The establishment of steel plants such as Bhilai and Bokaro in the 1950s and 1960s, and the Green Revolution, were foundational investments that reshaped the economy. Their legacy continues to influence employment, supply chains and regional development. Yet in recent decades, the economy has drifted towards the promise of financialization. Vertical industrial depth—where value is created across layers from design to manufacturing to branding—has given way to horizontal churn in financial markets. Financialization rewards capital gains and margins, benefiting a narrower class of participants, while marginalizing the labour, skill and innovation that once underpinned industrial growth.

Yet it would be both unfair and incomplete to deny the real benefits that financialization has brought to emerging economies like India. By deepening capital markets, it has expanded access to credit beyond traditional banking channels, unlocking funding for housing, infrastructure and enterprise. It has helped channel household savings into productive uses, moving capital from idle holdings into the formal economy. Instruments such as bonds and REITs have diversified risk and provided both governments and firms with additional tools to finance growth.

Financialization has also accelerated India's integration into global markets. The discipline imposed by quarterly disclosures, credit ratings and foreign investor scrutiny has pushed many firms towards improved governance and transparency. It has encouraged the professionalization of management and, in some sectors, shifted control from traditional promoters to institutional investors demanding efficiency and accountability. For a new generation of entrepreneurs, access to venture capital and private equity has reduced reliance on family wealth or state support, broadening the base of enterprise.

Yet it is precisely these gains that make financialization so seductive. Its speed conveys modernity, its logic appears mathematically precise, and its visibility—through rising indices and valuations—offers a constant signal of progress, even when underlying industrial capacity may not be expanding. For policymakers under pressure to demonstrate quick results, and for boards answerable to impatient investors, it can appear as an elegant shortcut: the promise of rapid wealth creation without the slower, more demanding process of building real productive capacity.

Freedom and Social Capital Are the Path Forward

India must take industrialization seriously—not merely as an economic choice, but as a matter of national resilience. A country of this scale and demographic complexity cannot rely indefinitely on a services-led model, even in peacetime. In an increasingly multipolar world, with competing trade blocs likely to emerge, manufacturing offers what few other sectors can: stable, large-scale employment; intergenerational technology transfer; and the creation of physical assets that deepen national capability. The ability to produce critical goods—from electronics to green energy systems—is not just prudent, but essential.

Industrialization also remains the most reliable pathway to move millions from informal, precarious livelihoods into organized, higher-value work. The ecosystems it creates—spanning suppliers, logistics, energy and skills—distribute gains more broadly across regions, rather than concentrating them in service-sector enclaves. Successful industrial economies demonstrate that factories do more than produce goods: they anchor communities and enable social mobility.

India's own experience reinforces this. Investments in steel, heavy engineering and fertilizers in the mid-20th century created the foundation for downstream industries and research capacity. The relative neglect of such long-horizon investments in recent decades has left gaps in critical sectors—from semiconductors to precision manufacturing—that services alone cannot fill. Renewed industrial ambition is therefore a necessity, not nostalgia.

Yet this pivot is inherently challenging. Services have expanded rapidly due to lower capital requirements, fewer regulatory frictions and faster returns. Manufacturing demands the opposite: high upfront investment, long gestation periods and coordinated policy across multiple levels of government. Sustaining such an approach in a system oriented towards short-term outcomes is difficult.

This is precisely why strong state leadership and investment are indispensable. Public capital must underwrite early risks in sectors such as advanced materials, electronics and strategic manufacturing—areas where private capital is often hesitant. As seen in East Asia, rapid industrialization requires a balance: private enterprise driving innovation, supported by public investment in infrastructure, research and demand creation.

History's lesson is clear. When finance becomes the master, industry becomes the victim and nations lose both resilience and relevance. If we want to build an economy that endures, we must put factories before future and options trading, engineers before traders and production before speculation. Only then can we claim growth that is not just swift, but sovereign.

Industrial capacity, however, depends on the successful implementation of market and societal reforms as a precondition for creating a broad-based economic foundation. Only when this base exists can innovation and growth in adjacent sectors take hold.

India's current position—particularly its limited leadership in critical digital technologies—reflects the absence of such a foundation. The country did not fully embrace the market and societal reforms required, even after liberalization. Moreover, the gains from these partial reforms have been uneven, benefiting only a small segment of the population. This has limited the creation of income surpluses and assets at scale, making it difficult for financial institutions to serve large sections of society with high-quality products and services.

This uneven distribution has also contributed to growing distrust between the state and citizens. The episode of the farm laws is instructive. In principle, allowing farmers to sell outside state-regulated mandis aimed to promote market efficiency and reduce state control—an objective not unlike China's early agricultural reforms. However, the process differed significantly:

- Chinese reforms maintained a dual system, where farmers could sell in open markets after meeting state quotas, effectively ensuring a price floor—something absent in India's case.
- Reforms in China were gradual, with pilot programmes and extensive consultation that built trust through demonstrated outcomes; the farm laws, by contrast, were introduced without sufficient consultation.
- As a result, Chinese farmers perceived reforms as empowering, while in India, they were seen as favouring markets without tangible, immediate benefits.

Societal support—built through incremental, participatory reform—is therefore essential for creating a durable industrial base.

The history of industrial development shows a consistent pattern: nations that lead globally control both the physical means of production and the knowledge embedded within them. In the 21st century, however, control over production alone is insufficient. Industrial capability without digital command—the ability to manage data, algorithms and software-defined systems—creates vulnerability. Equally, digital strength without domestic manufacturing depth is fragile.

India faces a subtle but profound dilemma: decades of uneven industrial development have created dependence on foreign inputs in critical sectors. A similar reliance exists in the digital domain, where essential cloud platforms, AI frameworks and semiconductor design tools are controlled by foreign companies. This dual dependency is perilous because the global system is increasingly shaped by nations willing to weaponize economic interdependence. In a world where export controls, sanctions and technological restrictions are increasingly deployed as instruments of power, this dependence is a strategic risk. Sovereignty in industry and in digital systems cannot be separated; each reinforces the other.

Both, in turn, depend on societal resilience. Policy can only succeed if it visibly improves livelihoods and quality of life. Without this, grand narratives—whether of a 'Digital India' or a US$5 trillion economy—risk losing credibility.

True sovereignty begins not in data centres or factories, but in the hidden bonds of trust that hold a society together. A nation cannot build resilient industries or digital strength on fractured ground. Without social capital—the trust between citizens, and between the people and the state—policy failure is inevitable.

Social capital is the foundation of freedom: the freedom to thrive, to collaborate and to build. In a society as diverse as India's, with its many caste communities (*jati*), trust deficit persists—not only between communities, but also between citizens and the state.

Only when policy and society consciously address this deficit can India forge a sovereign foundation strong enough to withstand global storms.

3

What Poor Customer Service in Indian Banks Tells Us about Industrial Inclusion

A cybercrime operator we once encountered online made an interesting confession: 'We are not a cottage industry, but an organized criminal enterprise. We have our structure everywhere, including within banks. How else do you think we move money at scale?'

This remark came during the height of the COVID-19 pandemic, when it was still unclear how funds from victims were being transferred so rapidly without detection. One telling anecdote, however, revealed how this nexus operated: many recipient bank accounts listed the bank's own address. In several cases, bank staff had opened these accounts in a single day, without even attempting to generate plausible fictitious addresses.

This pattern has continued. In one instance, Gurugram police arrested 27 bank employees within a single year for facilitating the opening of accounts under fictitious identities and renting them to scammers for a one-time fee of around ₹10,000 per account.*

Contrast this efficiency with the everyday customer experience. In one case, a startup attempting to open a current account at a private sector bank was turned away with hostility, including verbal abuse, with no meaningful redress despite filing a complaint.

*'Bank Employee among 2 Nabbed for Cyber Fraud in Gurugram', *The Tribune*, 22 February 2025, https://www.tribuneindia.com/news/haryana/bank-employee-among-2-nabbed-for-cyber-fraud-in-gurugram/. Accessed on 25 May 2026.

In another, a woman depositing cash was subjected to harassment and derogatory remarks* by a bank employee, who faced only minor disciplinary action in the form of withheld increments rather than dismissal.

Is this merely a one-off bad behaviour, or is there something more?

The RBI's Kanungo Committee report points to a systemic issue in customer service across Indian banks, with annual complaints exceeding one crore since 2019–20. More than half (54.5 per cent) relate to core banking functions: account operations (22 per cent), technology platforms (16.3 per cent) and cash management services (16.2 per cent).†

The prominence of technology being the Achilles heel of Indian banking‡ is particularly striking, given the common narrative of India's leadership in digital payments. Yet repeated outages because of regulatory action—such as the penalty imposed on HDFC Bank in December 2020 for prolonged service disruptions linked to chronic underinvestment in technology—indicate that these weaknesses are systemic and not confined to public sector institutions alone.

Walk into any branch of a large Indian bank today and the experience will feel familiar—but not necessarily reassuring. A security guard waves you in with polite indifference. You take a

---

*Das, Neelanjit, 'SBI Employee Penalised for Misbehavior: Bank's Disciplinary Action of Cutting Two Salary Increments Upheld by High Court', *Economic Times*, 14 June 2025, https://economictimes.indiatimes.com/wealth/legal/will/state-bank-of-india-sbi-employee-penalised-for-misbehavior-banks-disciplinary-action-of-cutting-2-salary-increments-upheld-by-high-court/articleshow/121774135.cms. Accessed on 25 May 2026.

†Reserve Bank of India Consumer Education and Protection Department, *Report of the Committee for Review of Customer Service Standards in RBI Regulated Entities*, Reserve Bank of India, https://rbidocs.rbi.org.in/rdocs/PublicationReport/Pdfs/REPORTCUSTOMERSERVICE0DC064A6613C40AB8343B94EF50DFA13.PDF. Accessed on 25 May 2026.

‡Mukherjee, Andy, 'Tech Is Indian Banks' Banana Peel. Everyone's Slipping', *Bloomberg*, 30 April 2024, https://www.bloomberg.com/opinion/articles/2024-04-29/tech-is-indian-banks-banana-peel-everyone-s-slipping. Accessed on 25 May 2026.

token and sit on a worn plastic chair, waiting to be summoned by a harried executive seated behind a low partition. If you're lucky, the air conditioning works; if you're luckier, someone can resolve your issue. More often, you are there because something has already gone wrong—a failed transaction, an unexplained charge or an errant auto-debit.

Now open your banking app. It is sleek, responsive and often feature-rich. You can transfer money instantly, apply for loans or invest in a mutual fund within minutes. It works well—until it doesn't. When something falls outside the script, support becomes elusive. Chatbots loop you through a maze of irrelevant responses and helplines test patience rather than provide resolution.

This is the paradox of modern Indian retail banking: polished on the surface, strained beneath. In a country where banking services are widespread, customers are often left navigating a system that feels impersonal, extractive and misaligned with their needs. Anecdotally, service quality appears to have deteriorated, suggesting the need for a renewed, updated assessment—perhaps a 'Kanungo Report 2.0' to capture the current state of customer services.

There was a time when 'relationship banking' was a promising idea—a strategy imported and adapted in the heady days of India's economic liberalization. In the late 1990s and early 2000s, as private banks emerged from the shadows of their lumbering public sector cousins, they introduced the language of customer intimacy. Relationship managers (RMs) were positioned as trusted financial advisers—attuned to a customer's life stage, aware of their goals and available to handhold them through complex financial decisions. For a while, the model worked. Home loans, auto loans and investment products were tailored to a rising class of salaried professionals. Branches were well-staffed with bright and suited young executives and service felt attentive. Banks welcomed you with a warm smile—and a freshly laminated credit card. There was even something aspirational about it all.

But slowly, the nature of this relationship shifted. Banks became more aggressive and RMs more transactional. Incentives changed, and cross-selling became the norm. Instead of understanding customer needs, managers were often driven to push products—such as Unit-Linked Insurance Plans (ULIPs) or health insurance plans—that had little to do with the customer's financial goals and everything to do with the manager's quarterly revenue targets.

Ask a banker today what 'relationship banking' means, and the answer is likely to sound scripted. Ask a customer, and the answer will feel like a sigh. There was a time when RMs stayed with customers for years—they knew the children's names, reminded them when a loan reset was due and helped renegotiate terms—building familiarity and continuity. Today, attrition rates in private banks hover between 25 and 30 per cent annually. Many RMs are young, undertrained and under pressure to meet targets. They last just long enough to pitch a product, then vanish. The human connection that once underpinned the model has been replaced by quotas and Key Performance Indicators (KPIs).*

Even high-income customers with premium accounts increasingly report confusion rather than confidence. Their airport lounge access may remain intact, but their financial journey is fragmented. Each year brings a new RM, a new mobile app update and a new push for products that often overlap, contradict or underdeliver.

As for middle-class and mass-market customers, there's little pretence left. They know the drill: show up, take a token, wait, argue, escalate, leave.

For all the rhetoric around high-touch service, the reality is far less flattering. Most so-called 'relationship managers' are glorified sales agents, evaluated not on the quality of relationships they build, but on the volume of products they sell. To the average

*KPIs are quantitative metrics used to assess individual or organizational performance against defined objectives.

customer, an RM is rarely a trusted adviser—more often, they are seen as someone trying to meet a quota, typically by pushing insurance policies or encouraging unnecessary portfolio churn. In some cases, the RM role exists less to serve the customer and more to satisfy the vanity of a certain class of clients who enjoy being 'assigned' a point of contact, even when that contact has limited ability to resolve issues.

Today, the real money in retail banking comes not from traditional products such as savings accounts or loans, but from the sale of non-banking products—insurance, mutual funds, credit card-related add-ons. This model, called bancassurance, is a financial goldmine. Unlike lending, which involves credit risk, capital provisioning and the cost of deposits, fee-based bancassurance generates immediate revenue with relatively limited side effects.

In FY2024 alone, bancassurance commissions contributed over ₹14,500 crore* to Indian banks. Private banks derive nearly 10 per cent of their fee income from these commissions, while public sector banks earn about 7 per cent. The pool is expanding rapidly—at an impressive 25 per cent CAGR—not necessarily because of customer demand, but because of sustained institutional push.

This pressure to cross-sell has predictable consequences. Mis-selling is rampant. Customers often discover recurring premium commitments only after funds begin to disappear from their accounts. Complaints have risen, prompting regulatory warnings, even occasional sanctions, from the RBI. Yet the underlying incentive structures haven't changed.

And why would they? As traditional lending margins come under pressure—thanks to rising deposit costs, regulatory constraints and increased competition—banks are compelled to

---

*'Banca Cap May Impact Banks' Net Profit by 1–2%, Says IIFL Securities', *Business Standard*, 20 December 2024, https://www.business-standard.com/finance/insurance/banca-cap-may-impact-banks-net-profit-by-1-2-says-iifl-securities-124122000849_1.html. Accessed on 25 May 2026.

mine existing customers for more revenue, even if it risks crossing into predatory or unethical practices.

The erosion of relationship banking might have been tolerable—even rational—if digital banking had seamlessly taken its place. But while India has made great strides in digital payments and infrastructure, the transformation of retail banking remains partial and uneven. Apps are designed for convenience, not guidance. Chatbots are quick but shallow. Personalization is still rudimentary—more pattern recognition than meaningful advice. And when something breaks—a KYC issue, a blocked card or a failed transaction—the journey from problem to resolution is often torturous.

To make matters worse, just as banks were losing their human touch, the customer base itself was undergoing a quiet transformation. The Indian retail banking customer is no longer what it once was. The average age is falling, digital fluency is rising and expectations are increasingly shaped by platforms outside finance—whether it is the ease of a Swiggy refund, the responsiveness of Amazon's support or the real-time tracking on Zomato. This is the benchmark banks are now expected to meet. Yet many remain in a reactive mode, layering technology onto legacy systems without reimagining what banking should feel like in the 2020s.

In the rush to digitize, banks have often embraced a shallow form of transformation—digitizing forms, automating branches and launching more apps and portals. These are not structural solutions but surface-level upgrades. What has emerged is the digital processing of old workflows, rather than a redesign of customer journeys.

Take, for instance, the now-ubiquitous 'video KYC'. In theory, it promises speed and convenience. In practice, it often demands stable high-speed internet, proper lighting, precise positioning, and multiple attempts to connect with an agent following a rigid script. This is not transformation so much as the relocation of friction from one channel to another.

Or consider grievance redressal. Many banks now require customers to log in, raise a ticket, select from predefined categories and then wait—often days—for a response that may not even resolve anything. The burden of navigation has shifted entirely to the user. The onus of explanation is placed on the already confused. And the system itself is designed around the assumption that problems are purely technical rather than recognizing the need for clarity, reassurance and resolution.

No wonder consumers are frustrated. They are told they are living in an era of digital transformation, yet what they encounter often feels like an old filing system wrapped in modern code—impersonal, opaque and impossible to engage with meaningfully.

For years, Indian banks could take their retail customers for granted. Switching was cumbersome, choices were limited and inertia—more than loyalty—was the glue holding relationships together. But that glue is coming undone. Younger customers, entering the financial system with digital-first expectations, carry neither the emotional attachment nor the practical constraint that shaped earlier generations. They are comfortable splitting their financial lives—maintaining a salary account with one bank, investing through a separate platform, using wallets for daily payments and turning to digital lenders for credit. Convenience matters more than legacy, simplicity trumps scale, and trust is no longer tied to brick-and-mortar branding—it is earned through experience.

Banks that still believe they enjoy lifetime customer stickiness risk misreading this shift. What they interpreted as loyalty was often just a lack of viable alternatives. But the moment those alternatives appeared—better designed, easier to access and more responsive—customers left without hesitation.

Fintechs—across payments apps, credit platforms, investment services or neo-banks—have not just introduced new products; they have rewired expectations. They have accustomed users to expect things that legacy banks have long resisted: instant onboarding, real-time updates, clear interfaces, contextual support and fewer hoops to jump through.

When a customer can open a savings account in a few clicks, get a card within days, track spending through intuitive dashboards and access real human support in-app—that became the new benchmark. Not a bonus, but a baseline.

The frictions banks normalized for decades—delays, excessive documentation, opacity—begin to appear untenable. Why should opening a fixed deposit take multiple visits and paperwork when a fintech can do it in minutes, with just OTPs? Why should updating a phone number require a branch visit when most digital platforms allow it instantly? This gap in experience is not trivial. It is tectonic.

Many fintechs have gained market share not by inventing radically new financial products, but by redesigning existing ones through a lens of user experience. They did not create mutual funds, gold loans or UPI. They just made them easier to understand, easier to trust and easier to use.

That said, in the enthusiasm of disruption, a sobering caution must be placed on the table. Fintechs do not operate outside the system; they are embedded within one of the most regulated and trust-sensitive sectors of the economy. Finance is not entertainment, and banking cannot be reduced to software layered with a pretty interface.

'Tech' in the name does not grant them a passport to bypass the discipline that governs deposit-taking, credit underwriting, data protection and consumer safety. Fintechs may innovate on the rails, but they cannot ignore the guardrails. Too many, in their early phases, have mistaken frictionless design for regulatory immunity. They speak the language of agility but overlook the grammar of accountability—prioritizing short-term growth over long-term stability. Some have operated credit products without appropriate licences;* others have treated customer data casually, harvesting and repurposing it behind opaque disclosures. This is not a disruption—it is a dilution of fiduciary responsibility.

If legacy banks were often careless with empathy, newer fintechs risk being careless with integrity.

At the heart of this lies the cultural core of the problem. Indian financial institutions, by and large, have not truly placed the customer at the centre of their design, delivery or dignity. The 'customer is king' mantra has remained a poster on the branch wall, not a principle embedded in behaviour. In practice, the customer is often treated as a data source, a target for cross-selling and an inconvenience when raising complaints. There are, of course, exceptions—thoughtful branches, responsible bankers and customer-first startups—but they remain exactly that: exceptions.

In most large banks, escalation remains the only effective recourse. Customers must write to nodal officers, approach the regulator, or name-and-shame on social media to be taken seriously. Basic actions—declining an unwanted product, reaching the same service agent or cancelling a credit card digitally—often require disproportionate effort. Meanwhile, banks retain the unilateral ability to auto-debit, upsell, auto-renew and send ten SMSs a day—all with impunity. Even the so-called 'high net-worth' clients who are courted with velvet gloves know the reality: if you want service, you must escalate; if you want a resolution, you must insist. And if you want clarity, you'd better do your own research.

This reflects a structural asymmetry of power, not merely isolated institutional failures. We propose that this is a structural design of the financial system and is not specific to a few banks. For example, only 50 per cent of the grievances turn into formal

---

*'Slice Temporarily Stops Users' Cards to Adhere to RBI's Digital Lending Rules', *Economic Times*, 29 October 2022, https://economictimes.indiatimes.com/tech/startups/slice-temporarily-stops-users-credit-cards-to-adhere-to-rbis-digital-lending-rules/articleshow/95169208.cms. Accessed on 25 May 2026.

complaints* (which are then surfaced in official publications such as the Kanungo Committee report) and as the table below indicates, fear of retribution, unlikely resolution and being advised not to pursue are significant indicators of the power hierarchy.

Product	*Did not complain*	*Fear of retribution*	*Resolution unlikely*	*Advised not to pursue by friends and family*	*Power asymmetry indication (per cent of the sum of all previous columns)*
Banking Deposit	1,048	38	188	19	23.37 per cent
Bank Credit	41	2	6	2	24.39 per cent
ATM Debit Card	758	17	120	17	20.31 per cent
Net Banking	251	7	43	12	24.70 per cent
UPI Wallets	344	5	60	15	23.25 per cent
NBFC	10	2	2		40 per cent
MFI	35	2	8		22.85 per cent
Co-op Credit Society	8	1			12.5 per cent

If the 50 per cent grievance-to-complaint ratio is extrapolated to the figures cited in the Kanungo Committee report, the actual number of grievances rises to approximately 2.22 crore in 2019–20, 2.02 crore in 2020–21 and 2.08 crore in 2021–22. This suggests that

*Balasubramaniam, Vimal, Aishwarya Gawali, Renuka Sane, and Srishti Sharma, 'Examining Grievances and Redress for Banking Products', *The Leap Blog*, 28 December 2022, https://blog.theleapjournal.org/2022/12/examining-grievances-and-redress-for.html. Accessed on 25 May 2026.

irrespective of socio-economic status, an individual is likely to face a serious banking-related grievance at least once every five or ten years,* with multiple incidents causing inconvenience being a given based on one's risk profile.

The Root(s) of Power Asymmetry

We present a set of interrelated factors that contribute to this structural imbalance:

a. FATF-led regulatory frameworks function, in part, as a global sanctions regime, encouraging nation-states to align their financial systems with broader geopolitical objectives. In practice, this often reinforces Western financial hegemony by enlisting other countries into compliance regimes that can be leveraged—alongside tariffs and export controls—to constrain geopolitical adversaries. When this approach—treating all as potential money launderers—intersects with a heavily licensed BFSI sector, institutions become more focused on regulatory compliance and risk avoidance than on customer service. The overriding incentive is to avoid actions that could invite regulatory sanction, rather than to optimize for customer experience.

b. Financial repression further shapes institutional behaviour. Banks are mandated to hold a significant proportion of their assets (much higher than the 10 per cent global norm) in government bonds under the Statutory Liquidity Ratio (SLR) requirement and other rules.† When only 5 per cent

---

*Recall from the first chapter that account being frozen because of KYC update is a phenomenon that will happen to everyone once in 2,5 or 10 years irrespective of their status.

†Chitgupi, Aneesha, Ajay Shah, Manish K. Singh, Susan Thomas, and Harsh Vardhan, 'Who Lends to the Indian State?', *Working Papers* No. 34, xKDR, 2024, https://papers.xkdr.org/papers/2024ChitgupiSSTV_who_lends_to_indian_state.pdf. Accessed on 25 May 2026.

of government borrowing is funded through voluntary market participation (mutual funds, private equity, venture capital and private individuals), it does raise questions about the government or the regulator requesting help in placing government debt on non-rule-of-law ways.*

c. The average annual income for the bottom 90 per cent of Indians is approximately ₹1.3 lakh,†‡ while a large proportion of the population (80 crore) depends on food security§ programmes such as the National Food Security Act, with a coverage gap of another 12 crore. This means most of the population does not have any surplus savings, constraining their ability to engage meaningfully with formal financial products; they can, hence, either reduce their consumption expenses to survive¶ or take loans to bridge the gap.** Nestle's Chair put it very bluntly, saying: 'There used to be a middle

---

*This refers to informal, extra-legal pressure. In practice, it often involves direct communication—such as phone calls—urging institutions or individuals to act outside the formal legal framework, sometimes accompanied by implicit coercion (for example, 'do it or face consequences').

†'Top 1% Indians' Share in National Income Is Higher Now than under British Rule: Data', *The Hindu*, 23 March 2024, https://www.thehindu.com/data/top-1-percent-indians-income-share-is-higher-now-than-under-british-rule-data-point/article67979494.ece. Accessed on 25 May 2026.

‡Bottom 50% = 50 people earning ₹71,000/year, Middle 40% = 40 people earning ₹165,000/year and Bottom 90% on average make ((50 * 71,000) + (40 * 165,000)) / 90 = ₹1,12,778.

§Paliath, Shreehari, 'Census Delays Leave 120 Million Indians Without Access to Foodgrain', *IndiaSpend*, 3 May 2025, https://www.indiaspend.com/governance/census-delays-leave-120-million-indians-without-access-to-foodgrain-951570. Accessed on 25 May 2026.

¶Bose, Shritama, 'Cracks in India's Consumption Story Run Deep', *Reuters Breakingviews*, 16 January 2025, https://www.reuters.com/breakingviews/cracks-indias-consumption-story-run-deep-2025-01-15/. Accessed on 25 May 2026.

**J.C., Anand, 'Savings Down, Financial Liabilities Up: RBI Says Household Debt Warrants Close Monitoring', *The Economic Times*, 27 June 2024, https://economictimes.indiatimes.com/news/economy/indicators/dipping-savings-increasing-financial-liabilities-mean-household-debt-needs-close-monitoring-rbi/articleshow/111313649.cms. Accessed on 25 May 2026.

segment, which used to be the segment that most of us fast moving consumer goods firms used to operate in, which is the middle class of the country, that seems to be shrinking'.

Taken together, these factors create a structural power imbalance in which the state holds more power than banks, and banks, in turn, hold significantly more power than individuals—because most of them have limited financial means. State-driven financial inclusion mandates, while well-intentioned, impose obligations on banks without necessarily making such accounts economically viable. As a result, banks often have little incentive to provide high-quality service to large segments of the population, and can sustain poor service standards with minimal consequences.

If one were to represent this hierarchy simply, it would be Government > Banks > People, with each layer exercising at least 100 times more power than the one below it.

Consider the case of Bank of Baroda, the second-largest public sector bank, which was caught in a scandal. The bank was found to have linked customers' accounts to mobile numbers belonging to bank officials,* without explicit permission, to meet internal targets. To understand how such an action is even possible, consider the following facts:

1. A bank account can be thought of as a document that is constantly updated and maintained by the bank automatically, but under a defined set of rules.
2. For instance, certain updates are automatic: if someone were to send money to an account, the balance is immediately updated to reflect the credit. However, withdrawals require explicit authorization depending upon the channel used: internet banking requires login credentials, mobile

---

*Gairola, Hemant, 'India's Bank of Baroda Tampered with Accounts to Flog App', *Al Jazeera*, 11 July 2023, https://www.aljazeera.com/economy/2023/7/11/indias-bank-of-baroda-misused-customer-data-to-flog-app. Accessed on 25 May 2026.

banking requires banking-registered device and PIN, and physical withdrawals require instruments such as ATM cards, cheques or withdrawal slips.

3. Hence, the bank, as an entity, always had unrestricted write access to the customer account at an entity level, but its credibility is established because this power is checked through regulatory oversight (by the RBI), internal controls (audits) and through a system of technical measures (product design).
4. Without this unrestricted write access to the customer's account, it is not possible for a bank to function because its business model is based on using the money put in by the depositor for lending to others at a higher interest rate to fund its operations and existence.

The abuse of such unrestricted write access—across hundreds of thousands of customer accounts—to satisfy internal targets set by the management reflects a deliberate choice to monetize institutional power for short-term gains that may not even result in long-term additional income. The method employed also reveals the underlying power dynamics. Accounts without linked mobile numbers were specifically targeted, on the assumption that such customers were less likely to detect or challenge irregularities because they were typically among the most economically vulnerable.

That this made these accounts easy to take over is clear from Bank of Baroda's own communication as explained below:[50]

1. Users download the bank's mobile app
2. They enter the mobile number linked to their account
3. An OTP is sent to this number for verification
4. The device is authenticated through a silent SMS sent from the SIM card to the bank's servers
5. Once verified, the device is bound to the account and a six-digit PIN is set to enable transactions

While similar steps are required to update an existing mobile number, accounts that were already linked were avoided, as any change would have immediately alerted the account holder and triggered complaints. Instead, unlinked accounts—effectively unmonitored—were chosen.

This power dynamic (of being poor) being the root cause of the scandal is clear from the investigation. An officer from Rajasthan, speaking anonymously, described a similar gimmick to *Al Jazeera*: he said his branch launched a campaign to open zero-balance accounts to attract unskilled labourers and daily-wage workers, enrolling them into digital platforms without meaningful consent. He added that the staff did explain the linkage between the app and their money, and readily uninstalled it for those who were wary. The officer noted the irony of enabling digital banking for those who barely make ends meet: 'At the end of the day, to meet the number [target] and save your bread and butter, you have to do such things.'*

For any bank, irrespective of its size and profile, there is a fixed cost associated with servicing a bank account. Industry insiders estimate that maintaining a zero-balance account costs roughly ₹500 annually in software and hardware costs alone, rising to around ₹1,500 when branch infrastructure and staffing costs are included. With an average net interest margin (NIM) of about 3.5 per cent† for Indian banks, a bank would need an average balance of approximately ₹42,857 (₹1,500/0.035) to break even. However, not all public sector banks have the same NIM: State Bank of India's NIM is 2.86 per cent, Bank of Baroda's is 2.93 per cent and Punjab National Bank's (PNB) is 2.67 per cent.‡ Lower NIMs translate

---

*Ibid.

†'Indian Banks' Margins to Take 10 Bps Hit in FY 26 Due to Rate Cuts, Fitch Estimates', *Reuters*, 13 February 2025, https://www.reuters.com/business/finance/indian-banks-margins-take-10-bps-hit-fy-26-due-rate-cuts-fitch-estimates-2025-02-13. Accessed on 25 May 2026.

‡Singh, Yuvraj, and Beenish Bashir, 'Indian Banks Face Renewed Pressure on Net Interest Margins as Rates Head South', *S&P Global*, 19 February 2025, https://www.spglobal.com/

into a higher average balance of ₹52,447, ₹51,945 and ₹56,179, respectively.

In reality, these estimates are conservative. When additional factors—such as the Aadhaar Enabled Payment System (AePS) interoperability, dormant accounts, prohibitive KYC mandate and the cost of issuing debit cards—are accounted for, the annual servicing cost can rise to around ₹3,500.* This pushes the break-even balance requirement to ₹98,511 (average), ₹1,20,628 (SBI), ₹1,19,473 (BOB), and ₹1,29,211 (PNB). However, most accounts do not have a balance of more than ₹6,000 on average, which means that banks almost always incur losses on these accounts and would have little incentive to maintain or open them. The power differential between banks and the government, however, coerces them to open such accounts and, in response, banks seek to recover these costs indirectly—often through service fees or cross-selling.

For instance, public sector banks collected more than ₹15,000 crore† in penalties for non-maintenance of minimum balance between 2014–15 and 2023–2024. At the same time, the growth of mobile banking has sharply increased the volume of SMS alerts mandated by RBI regulations for both debit and credit transactions. For FY2024–25, the total number of such alerts could be as high as 38,28,380 crore‡ (twice the volume of IMPS and UPI transactions).

---

market-intelligence/en/news-insights/articles/2025/2/indian-banks-face-renewed-pressure-on-net-interest-margins-as-rates-head-south-87612169. Accessed on 25 May 2026.

*Nayak, Siddhi, 'Banks Struggle to Make Jan-Dhan Accounts Cost Effective and Viable, Eight Years On', *Moneycontrol*, 2 June 2022, https://www.moneycontrol.com/news/business/eight-years-on-banks-struggle-to-make-jan-dhan-accounts-cost-effective-viable-8621681.html. Accessed on 25 May 2026.

†Muthyanolla, Sai Krishna, 'Data: Between 2014–15 and 2023–24, PSBs Collected Over ₹15,000 Crores in Penalties for Failing to Maintain Minimum Balances', *Factly*, 9 August 2024, https://factly.in/data-between-2014-15-and-2023-24-psbs-collected-over-%E2%82%B915000-crores-in-penalties-for-failing-to-maintain-minimum-balances. Accessed on 25 May 2026.

‡Reserve Bank of India, *RBI's Annual Report*, p. 165, 'Table IX.1: Payment System Indicators – Annual Turnover (April–March)', https://rbidocs.rbi.org.in/rdocs/AnnualReport/

Even at an extremely low pricing (1–5p per SMS), the cost would be between ₹38,283 crore and ₹1,91,419 crore, which can only be recovered through either a fixed usage charge or a per-alert usage fee. Requests to shift these notifications to app-based alerts* have, so far, not resulted in the relaxation of these regulations.

The roots of this power asymmetry between the government and banks can be traced to the era of bank nationalization. It continues to manifest in what former RBI Governor Dr Urjit Patel has described as the 'fiscalization'[†] of the banking sector—where banks are used as instruments to support broader fiscal objectives, particularly when direct government capacity is constrained.[‡]

For instance, in the push for Direct Benefit Transfer, public sector banks were given explicit targets for account opening, and the progress was meticulously tracked via monthly meetings. Banks only had to meet these targets and then be off the hook, without bothering about customer service at all[§], because their goal was never customer satisfaction or relationship but appeasing the government.

This model of directing banks to achieve government objectives is not restricted to public sector banks but extends to the entire banking system via formal mechanisms such as State Level Bankers' Committees. Meeting minutes[¶] indicate that even private sector banks' Annual Credit Plans (ACPs) are directed by the government,

---

PDFs/09PAYMENT29052025564CFE8E29164796AC1B2016508B51A6.PDF. Accessed on 25 May 2026.

*Shetty, Mayur, 'Allow App Pings instead of SMS: Banks', *Times of India*, 16 June 2022, https://timesofindia.indiatimes.com/business/india-business/allow-app-pings-instead-of-sms-banks/articleshow/92239290.cms. Accessed on 25 May 2026.

†Patel, Urjit, *Overdraft: Saving the Indian Saver*, HarperCollins, Gurgaon, 2020.

‡Rai, Suyash, 'Economic Development and Digital Transformation: Learning from the Experience of Aadhaar and Financial Inclusion in India', *XKDR Forum Working Papers*, No. 35, xKDR, 2024.

§Ibid, p. 28.

¶State Level Bankers' Committee Karnataka, *Proceedings of the 167th SLBC Meeting Held on 26.11.2024*, Bengaluru, https://archive.org/details/167th-slbc-meeting-proceedings-copy. Accessed on 25 May 2026.

despite these being, in principle, acknowledged as the prerogative* of each bank's independent business strategy.

When banks do not meet the targets assigned by the government, coercive actions are not uncommon. Instances of FIRs being filed against branch managers† have been the trend for a decade;‡ in some cases, sustained work pressure also led to severe stress among bank staff, leading them to even commit suicide.§

Financial Inclusion Requires a Base

Financial inclusion is defined by the RBI as access to financial services, including credit, to vulnerable groups at an affordable cost. However, this remains a monetary definition rather than a fiscal definition, as it explicitly leaves out income generation, which is a prerequisite for survival. If there is no stable or sufficient income generation, access to financial services alone cannot ensure inclusion. Where income is irregular or concentrated in short bursts, households are unable to smooth consumption effectively, often leading to taking debt to bridge gaps.

Rural household free cash flow (income minus expenditure) had remained largely stagnant—rising only marginally from ₹1,413 in 2016–17 to ₹1,436 in 2021–22.¶ With inflation increasing over the same period, household indebtedness has risen from 47.4 per cent to 52 per cent. As income opportunities and job creation slow down,

*Ibid, p. 6.

†State Level Bankers' Committee, Maharashtra State, *Minutes of the 164th SLBC Meeting Held on 29 August 2024*, p. 17, https://archive.org/details/slbc-maha. Accessed on 25 May 2026.

‡Kapre, Shubangi, 'FIR Lodged against 16 Banks for Failing to Give Crop Loans', *The Indian Express*, 18 July 2015, https://indianexpress.com/article/cities/mumbai/fir-lodged-against-16-banks-for-failing-to-give-crop-loans. Accessed on 25 May 2026.

§'Chief Manager Dies by Suicide in Bank of Baroda in Baramati Due to "Work Pressure"', *The Indian Express*, https://indianexpress.com/article/cities/pune/chief-manager-suicide-bank-of-baroda-baramati-work-pressure-10135199/. Accessed on 25 May 2026.

¶Arman Financial Services Ltd., *Annual Report 2024–25*, https://armanindia.com/pdf/AnnualReportPDF/pdf_718ff996-9a50-4c70-9bd6-e12c6616095b.pdf. Accessed on 25 May 2026.

the viability of finance itself collapses. This is because credit, by design, only has an amplification effect—it magnifies either surplus or deficit, but the multiplier itself cannot expand indefinitely. As it approaches a limit, debt servicing begins to erode the already limited free cash flow available for productive use or consumption.

The Joint Liability Group (JLG) model functions effectively only when there is a balance between borrowers and lenders. When too many debtors chase the same demographic, credit may initially expand, but the JLG model becomes unstable as collective indebtedness rises. While market reforms in agriculture and industry can (in theory) drive up income levels and create sustainable surpluses, India's growth trajectory has leaned more towards services and consumption. These sectors have not been sufficiently broad to generate consistent income growth for rural households.

Hence, the structural challenges within financial services act as a foundational constraint on India's ability to assert industrial and digital sovereignty. Financial inclusion, when pursued without corresponding growth in industrial and income participation, produces a superficial expansion: millions of accounts exist, yet large segments of the population remain economically constrained, unable to generate the surplus necessary for entrepreneurship, technological adoption or innovation. In the 21st century, sovereignty rests not on paper accounts, but on the capacity of citizens and institutions to participate meaningfully in productive and digital ecosystems.

The risks of neglecting these structural gaps are multi-layered. Economically, reliance on foreign platforms and intermediaries persists, as domestic actors struggle to scale without adequate capital or technological literacy. Strategically, delays in domestic development in key sectors deepen external dependency, limiting India's ability to shape global standards or compete effectively. Socially, friction-laden financial systems erode trust, weakening citizen engagement with state-led initiatives and slowing the emergence of a robust entrepreneurial class.

Financial inclusion, in isolation, cannot create a resilient economic foundation. Without parallel industrial and income empowerment, households and small businesses remain constrained by low earnings and poor access to capital to meaningfully participate in digital ecosystems or adopt emerging technologies. This creates a dual vulnerability: the population is economically limited, and the nation's technological ambitions remain dependent on external providers. Without strengthening domestic capacity to innovate and produce, India risks permanent dependency in strategically vital areas such as AI, cloud infrastructure and semiconductors.

The connection between financial inclusion and sovereignty becomes even clearer when viewed through the lens of time-sensitive technological change. Unlike traditional industrial expansion, breakthroughs in AI and digital platforms unfold over months, not decades. Delay or underinvestment in building domestic financial capacity directly slows the development of entrepreneurial and technical talent that drives such innovation. Every unbanked, underserved or financially constrained citizen represents not only a missed market opportunity, but also a gap in the nation's ability to build the human capital required for industrial and digital leadership. The cost of inaction is therefore not merely social or economic—it is strategic.

Finally, building sovereignty thus requires an integrated approach that connects financial access with industrial participation, skill development and digital infrastructure. Policy cannot stop at opening bank accounts; it must ensure that these translate into productive use, technology adoption and entrepreneurial activity. Only by creating an ecosystem where citizens are both financially empowered and technologically enabled can inclusion evolve into capability—turning access into a driver of domestic innovation, industrial growth and, ultimately, strategic independence.

The stakes are high: without such integration, India risks remaining a passive consumer in the very sectors that define sovereignty, ceding power and influence to nations that have successfully aligned financial empowerment with industrial and digital capacity. This structural context makes it clear why financial inclusion, industrial capability and digital sovereignty are inseparable.

We conclude with a simple hypothesis: industrial inclusion (and by extension, income inclusion) is a necessary precondition for meaningful financial inclusion. Access to a bank account does not, by itself, generate income surplus at an individual level. When 90 per cent of the population earns less than ₹2.9 lakh per year, but the cost of servicing an account is around ₹3,500, poor customer service for a large segment of the population becomes a default.

Low digital literacy and limited income, coupled with pressure on banks to meet aggressive targets, create incentives that can push not only individual bank employees but also, at times, portions of the bank to go rogue, as illustrated by the Bank of Baroda case.

While the Indian state's push for financial inclusion is commendable, both the mechanisms employed and the absence of prior industrial and income inclusion have created distortions. These side effects are not fixable because they remain rooted in a command-and-control framework from the 1960s.

This, we argue, is the primary reason why India has not yet produced breakthrough capabilities in foundational digital technologies such as LLMs. For a country to emerge as a hub in such domains, it typically requires at least one of three strengths: industrial power (that generates a lot of surplus), financial dominance (that confers extraordinary currency advantages) or a strong research ecosystem (that participates in the global technological supply chain). India, lacking sufficient scale in any of these, has faced structural constraints. While this does not preclude the emergence of competitive capabilities, it makes such outcomes less probable—positioning India, for now, more as a consumer (of advanced technologies) than a creator (of fundamental building blocks).

4

The Inter-Connected Industrial, Digital and Financial Supply Chain

Our parents were born in the immediate aftermath of Independence, when queuing for essentials at ration shops was a part of everyday life. In those days, India simply did not produce enough food to feed its population and depended on imports until the Green Revolution of the 1970s. Even in our own childhood, standing in queues, be it for food grains or at the post office to pay annual stamp fees for a transistor radio, was not uncommon.

What then explains the recurring shortages of fertilizers today in many parts of the country, forcing farmers to stand in line for hours just for a single bag? Several causes exist, but a few stand out:*

- India remains heavily dependent on fertilizer imports. Disruptions caused by conflicts such as the Russia–Ukraine war have affected supply chains, increasing costs and delivery times.
- Most of the imports are, however, from China, which has, at times, imposed unofficial restrictions following geopolitical tension—such as the three-day stand-off weapons war between India and Pakistan, monikered 'Operation Sindoor'.

*Pandey, Ashish, 'Shortage of Key Fertiliser Exposes Indian Agriculture's Alarming Dependency on China Imports', *Economic Times*, 15 December 2024, https://economictimes.indiatimes.com/small-biz/trade/exports/insights/shortage-of-key-fertiliser-exposes-indian-agricultures-alarming-dependency-on-china-imports/articleshow/116193909.cms. Accessed on 25 May 2026.

- These supply constraints create a classic bottleneck effect: a large number of farmers competing for limited stock. This, in turn, fuels hoarding, price spikes and diversion of supplies.

Rising fertilizer prices increase input costs, and it is only a matter of time before this feeds into higher prices for staples and other agricultural products for retail customers. Yet, given the structure of Indian agriculture, farmers will barely be able to recoup their costs, leaving them further entrenched in debt. And that means, consumption will stall, microfinance and rural debt agencies will struggle to collect payments, and this may eventually result in a farm loan relief scheme a few months down the road.

Supply chains, thus, not only transcend sectors but also impact them over time through cascading effects. The primary reason is that the economy is not composed of linear, self-contained industries operating in isolation. Rather, it is a web of interdependent relationships—all connected by thin, gossamer threads—where one industry's output becomes the input of another, and disruptions in one node ripple across the entire system.

In that sense, it resembles the human brain: several billion neurons connected to each other in myriad ways. However, in the case of the economy, these 'neurons' are not connected by dopamine (as in the brain) but by an idea called money. The management of this 'dopamine' is the role of central banks and governments. Just as flooding the brain with dopamine through drugs creates addiction—rendering a person incapable of productive activity—over-financialization, similarly, prevents the economic web from expanding (where expansion means the creation of more products and services).

Modern digital technologies, however, have transformed the nature of money. It is no longer primarily physical but increasingly digital. In the past, 'printing money' meant running presses to produce currency notes; today, it typically involves creating accounting entries in bank accounts. Moving money between accounts is essentially a process of database reconciliation over time, governed by messaging systems and well-defined ledger-management procedures. Central and commercial banks have been performing such functions for over four decades in different forms.

The key difference introduced by instant payment systems is the presence of multiple participants—consumers, Big Tech firms, financial intermediaries, fintech companies and payment gateways—and the fact that reconciliation now occurs within a 10–15 minute window across all transacting entities, rather than at end of day as was the case earlier. Payment technologies are digital, but they are distinct from other forms of digital infrastructure and fundamentally different from industrial supply chains.

One way to understand this distinction is to think of digital payment technologies as analogous to the body's mechanisms for producing and regulating dopamine. Efficient management of dopamine does not, by itself, make an individual proficient in any domain; that requires separate capabilities, skills and sustained practice. Similarly, efficient payment systems do not automatically translate into broader technological or industrial capabilities.

Thus, while the absence of domestic payment technology is a strategic vulnerability, its presence does not imply that a country possesses the capacity, skills, training or resources to develop other digital or industrial technologies. In simple terms, digital systems—particularly payment technologies—enable financial and industrial supply chains to organize, move faster and become more efficient. However, they do not substitute for the underlying capabilities required to build those systems. These domains are interconnected, but they remain distinct.

The Everlasting Big-Peace, Small-Wars Assumption

Much of the world order today is built on a hidden assumption—that there will be a unipolar world in which the question of who is in charge is largely settled. While conflicts may arise, they are expected to be contained and managed, because the hegemony of the US is assumed to endure and is implicitly baked into the system. India's foreign policy from the 1990s onwards largely followed this assumption, and was richly rewarded for doing so.

As the US opened its doors to India and Indians after the nuclear sanctions, both the country and its people benefited. A wealthy middle class emerged through the services industry and its savings were channelled into more productive enterprises. GDP and growth rates—often the true currency of power in the world—expanded significantly.

There was little perceived need to go back and fix industrialization, which the country had effectively skipped, because it was widely believed that India had leapfrogged into the digital era. This belief rests on a flawed mental construct—that the supply chains connecting industrial, digital and financial ecosystems are not only distinct, but largely independent and capable of existing on their own. As China demonstrated during the period following Operation Sindoor—through its unofficial restrictions on fertilizers and rare-earth magnets—it is possible to prevail in a limited military standoff and yet incur significant economic and strategic costs.

After all, what use is military capability if food supply chains—dependent on fertilizers—can be disrupted at will, not by an adversary like Pakistan, but by a strategic partner such as China?

In the modern digital–industrial–financial era, sovereignty is no longer defined solely by borders or conventional military strength. It increasingly depends on the capacity to generate, control and deploy industrial, digital and financial systems at scale. Nowhere is this more evident than in the domain of LLMs. The LLM supply chain—spanning datasets, model architectures, software frameworks, finance, cloud compute and physical infrastructure—is not merely a technical ecosystem; it is a strategic architecture of power. Control over this chain determines which nations can independently shape AI systems, set digital norms and safeguard critical knowledge infrastructure. Each layer represents both a potential vulnerability when externally controlled and a source of strength when domestically anchored. Sovereignty in the 21st century is therefore inseparable from mastery over this supply chain.

The ideological significance of this mastery is profound. LLMs do not merely process information; they encode compressed representations of human knowledge, embedding assumptions, biases and frameworks for reasoning. When their supply chains are externally sourced, the epistemic foundations of domestic decision-making—across finance, governance, healthcare and education—are partially shaped by foreign actors. Dependence on externally developed models is not merely a technological risk; it is a structural exposure. True sovereignty requires the ability to produce, validate and deploy models aligned with local realities, norms and values.

The LLM supply chain also illustrates how industrial, financial and research power converge to create asymmetric advantages. Nations that dominate multiple layers—compute infrastructure, finance, industry, core software frameworks, model code and datasets—emerge as hubs of AI innovation and influence. The geography of AI clusters reflects this concentration: those who control GPUs, cloud infrastructure and training capacity effectively shape the flow of knowledge, determine access costs and influence global adoption. Countries lacking such control remain dependent on decisions made elsewhere, limiting their ability to exercise independent agency.

Financial power is especially decisive. Training and maintaining foundation models require enormous capital—often running into billions of dollars. Without access to such resources, meaningful participation in model development is not possible. Digital sovereignty, therefore, is inseparable from the ability to mobilize capital—whether to acquire compute, attract talent or sustain infrastructure. Intellectual and industrial capacity alone is insufficient.

Finally, the case for sovereign independence in AI extends beyond strategy into ideology. The ability to generate and control LLMs is a marker of self-reliance in the knowledge economy. It enables nations to define their own epistemic frameworks, technological priorities and societal safeguards. It is about retaining the freedom to innovate, adapt models to local languages and cultures, and protect data and insights from external control. In this sense, the LLM supply chain is not merely technical—it is a blueprint for digital resilience and intellectual self-determination.

For India, the current landscape reflects both promise and fragility. While the country has a strong talent pool and a vibrant startup ecosystem capable of adapting and deploying global LLMs, the upstream layers—original architecture design, large-scale pretraining and unrestricted access to advanced compute—remain largely inaccessible. This creates a structural asymmetry: India can participate in applications, but cannot independently shape the evolution, standards or alignment of these systems.

This asymmetry is not just technical; it is ideological. A nation's ability to innovate in LLMs reflects its capacity to encode, interpret and deploy knowledge on its own terms. Without control over the intellectual, financial and industrial levers of LLM production, digital narratives—from language processing to policy guidance—are shaped externally.

As of the time of writing, the following foundation models are known to exist:

Model Name	*Developer/ Organization*	*Country of origin*
GPT-4o, GPT-4.5, Grok 3	OpenAI/xAI	United States
Claude 4 (Opus & Sonnet)	Anthropic	United States
Gemini 2.5 Pro, Gemini 1.5	Google DeepMind/ Google	United states

Model Name	*Developer/ Organization*	*Country of origin*
Llama 3 (including 3.1, 3.2, 3.3)	Meta AI	United States
Mistral (Mixtral, Small, Medium)	Mistral AI	France
Falcon 40B/180B	Technology Innovation Institute	United Arab Emirates
DBRX	MosaicML/ Databricks	United States
DeepSeek R1/V3	DeepSeek AI	China
Qwen 3, Qwen 2.5 Omni	Alibaba Cloud	China
Doubao 1.5 Pro	ByteDance	China
Ernie Bot/Ernie 4.5	Baidu	China
Hunyuan (Yuanbao)	Tencent	China
GLM-4.5	Zhipu AI	China
Baichuan 2	Baichuan AI	China
Velvet 14B, Velvet 2B	Almawave (Velvet AI)	Italy

The country of origin is significant because each has taken a distinct path in building an ecosystem where developing large-scale foundation models became feasible. Before examining those pathways, however, a brief overview of AI itself is necessary.

Knowledge as Compression

When children enter the third or fourth grade in primary school, they are introduced to multiplication tables. The method used is repetition—reciting them aloud in school and at home—until they can recall tables from 1 to 15 from memory. However, beyond that stage, we do not expect them to memorize all tables. Instead, we teach them general rules of multiplication, such as the carry method and shortcuts for specific numbers like 5, 10, 20 and 100.

If one were to write down all these rules, they would occupy no more than 15 pages. In contrast, writing out multiplication tables for all numbers, up to even four digits, would run into several thousand feet of paper. Using rules is therefore a far more efficient way of representing knowledge than memorizing every possible output. In other words, a higher-order rule that captures a concept compresses knowledge and is far more memory-efficient than storing all its outcomes.

As human beings, we constantly rely on such compression of knowledge to navigate the world. This has interesting side effects—something that, incidentally, was recognized even in ancient traditions.

Consider the story of Yavakrita in the *Mahabharata*. Jealous of the fame his peers enjoyed for their knowledge (then represented by the Vedas and Shastras), he sought to acquire all knowledge rapidly through penance directed at Indra, the king of the heavens. Indra, though impressed, warned him: how could one hope to hold the entirety of knowledge without the discipline required to understand it in depth? Without such discipline, one would be prone to grave errors of judgement.

In modern terms, Indra was warning Yavakrita about the side effects of compression. A general rule may work well in most situations and be more efficient than memorizing every detail, but it inevitably involves information loss. This loss can lead to serious errors, especially when past success creates overconfidence in one's abilities derived through compression of knowledge.

Lord Indra's warning proved correct. Yavakrita* eventually died

*The story of Yavakrita is much more complex than paraphrased here. But since we are discussing about knowledge as compression, it was deemed sufficient to compress that story to illustrate the side effects of too much compression resulting in the death of its practitioner.

at the hands of a demon because he lacked the practical knowledge required to defend himself—specifically, he did not carry enough holy water to chase the demon away while chanting mantras and holding the water in one hand.

The Tower of Babel offers a different but related perspective. It describes an ancient civilization that spoke a single language and attempted to build a tower reaching the heavens. Fearing their prowess, the gods fragmented their language, making mutual understanding impossible and halting the project. Humanity, thus, ended up being divided into multiple languages, tribes and states.

One way to interpret this myth in the context of knowledge as compression is through Algorithmic Information Theory (AIT). Consider the fraction one-third: it represents an infinite sequence of 0.333…, yet can be represented succinctly. Similarly, a sequence like '10101010101010' can be represented compactly as '10 × 7'. A smaller representation can encode a much larger sequence.*

Knowledge, when compressed, thus offers several crucial advantages:

1. It is easier to transmit, as it is more efficient.
2. It creates a shared vocabulary and structured ways of applying knowledge to real-world situations.

---

*This is called Kolmogorov complexity, a way of measuring how simple or complex something is by asking: What is the shortest possible description of it? Formally, the Kolmogorov complexity of a string is the length of the shortest computer program (in some fixed programming language) that can produce that string as output.

Examples:

AAAAAA
→ Can be described as: 'print "A" six times'
→ Low Kolmogorov complexity
ABABABABAB
→ Can be described as: 'print "AB" five times'
→ Still relatively low complexity
X7q$L9mP
→ No obvious pattern exists
→ The shortest description is likely the string itself
→ High Kolmogorov complexity

3. Repeated cycles of compression and decompression enable faster generation of new knowledge, as ideas can combine and evolve.

History illustrates the first point clearly. The advent of writing materials such as papyrus or palm leaves made knowledge transmission far more durable than oral traditions. The second point is obvious when we consider a spoken or written language across any period—all had a sentence structure (subject, predicate), vocabulary (a continuously evolving list of words), categorization of words (e.g., noun, adjective) and rules governing the word category in the sentence structure.

Scientific progression proves the third point. Our understanding of the physical world has accelerated dramatically, in part because of systematic ways of applying and refining compressed knowledge. Two principles are central:

- As Karl Popper argued, the scientific method disproves hypotheses with certainty but treats proof as conditional, based on existing evidence.
- When evidence contradicts an application of decompressed knowledge, then the specific decompression is considered flawed but not the decompression itself.

Hence, one way to think about the novel application of a decompression rule is that it can potentially generate new knowledge. Objects had been falling to the ground for centuries, but Galileo Galilei demonstrated that they fall at the same rate regardless of weight. Newton, building on this insight, applied the knowledge to understand why objects always fall in a straight line and formulated the theory of gravitation, expressing it mathematically through the inverse-square law.

Mathematics itself is a specialized form of knowledge compression, much like language. It has its own vocabulary, rules, structure and semantics. However, compression differs across domains. Over time, specialized fields develop their own compressed representations, which can become so efficient and domain-specific that they are difficult for others to understand.

In this sense, knowledge fragments—much like in the Tower of Babel. Specialists may struggle to communicate across domains because their compressed frameworks are not mutually intelligible.

True insight often emerges when these different forms of compressed knowledge interact. A useful analogy is biological reproduction. The genetic information of a species is compressed within DNA. Reproduction decompresses and recombines this information from two parents, creating variation. The survival of the new organism depends on how well it adapts to its environment.

Creativity—whether biological or intellectual—can thus be understood as an emergent phenomenon arising from repeated cycles of compression and decompression within specific contexts. It may even be formalized in terms analogous to Kolmogorov complexity.

Hallucination Is a Structural Side Effect

Constant cycles of compression and decompression, when applied at scale across varied situations, either succeed or fail. Consider Newton's insights on gravity or Einstein's theory of relativity. At the time of their conception, they were merely novel ideas without supporting evidence. Until such evidence emerged, they had no empirical basis. In that sense, they did not 'exist' as validated knowledge and were indistinguishable from what we might call a hallucination.

The distinction between a novel creative insight and a hallucination is therefore not evident at the moment of conception. It only becomes clear over time, as evidence accumulates and either supports or disproves the idea—consistent with Karl Popper's view of the scientific method. When evidence is absent or inconclusive, it is difficult to determine whether an idea is a future insight awaiting validation or merely an unfounded conjecture.

Philosophically, hallucination—perceiving what does not exist—becomes more likely as layers of compression increase. Consider reality itself. As Lord Indra warned Yavakrita: 'How do you expect to hold all knowledge of reality within your small mind, when your senses constantly mislead you?' While ancient thinkers did not describe this in terms of 'lossy compression', the insight is similar. Our sensory systems capture only a small fraction of the world due to inherent bandwidth limitations, filtering out information. In Indian philosophy, this is captured by the concept of Maya—the idea that the world perceived through the senses is not reality itself, but a constructed approximation. Visitors to India often remark on sensory overload, which is itself a reflection of this filtering process.

Sensory information is itself compressed before it reaches the brain, which then performs further compression. For example, visual processing involves multiple stages of compression* from the eye to the brain: signals from approximately 120 million photoreceptors are reduced to about 10 million bipolar cells, and then to around 1 million ganglion cells before reaching the brain. The brain then reconstructs the image from this compressed input. Blind spots are an inevitable consequence of this design, as the optic nerve creates gaps in the visual field. The brain compensates by filling in missing information, and when this process exceeds certain limits, hallucinations or visual distortions can occur.

Once we accept that repeated lossy compression can lead to hallucination, consider human language itself. It is widely acknowledged that there is always a gap between what is understood and what is expressed in writing. This gap is another form of lossy compression. Regardless of the language used, written text represents only a partial model of reality. Even complete mastery of a language does not equate to a complete understanding of the world—it remains a partial representation.

*It is a very different type of compression unlike JPEG, but it is compression, nevertheless.

Large Language Models as Lossy Compression of (Several) Partial World Models

What if there existed a machine that could compress all written knowledge across languages and then decompress it to answer specific questions? Such a system would act as a highly efficient semantic compressor—retaining the essence of knowledge and reconstructing context-specific responses when needed. It would be akin to distilling an ocean of information into its core patterns and regenerating relevant portions on demand.

An LLM is a concrete implementation of this idea. It does not compress text in the conventional sense (as ZIP algorithms* do), nor does it store facts in a structured database. Instead, it encodes knowledge as statistical patterns that enable it to generate plausible responses.

This is achieved through neural networks composed of layers of artificial neurons, analogous to biological neural structures. Each connection between neurons has an associated weight—a numerical value that determines how strongly one neuron influences another.

During training, the model is exposed to vast amounts of text and learns to predict the next word in a sequence. With each prediction, it adjusts its internal weights to improve accuracy. Over time, billions of such weights encode patterns related to language, grammar, factual associations and even elements of reasoning. These weights collectively form a lossy, compressed representation of the data the model has been trained on.

LLMs thus represent an additional layer of lossy compression applied to human language, which is itself a compressed representation of reality. One way to visualize this is as a large but

---

*'Zip algorithms' generally refer to compression algorithms used in formats such as ZIP. Their purpose is straightforward in principle: to reduce the size of data by identifying and removing redundancy. Instead of storing repeated patterns multiple times, the algorithm stores a shorter description of those patterns, allowing the original data to be reconstructed later.

slightly blurred picture of the world—generally accurate in broad structure, but lacking precise detail. When asked a specific question, the model does not retrieve exact stored knowledge; instead, it reconstructs an answer by approximating the most likely response from surrounding patterns.

In this sense, LLMs function similarly to human vision. When focusing on a particular detail within a larger image, the brain fills in missing information based on context. Most of the time, this approximation is sufficient. However, when the extrapolation exceeds certain limits, it results in optical illusions. LLMs operate in an analogous manner and are therefore inherently prone to hallucination.

The Supply Chain for the LLM

The basic requirements for creating any product are inputs, labour and capital. Consider a bag of 100 small screws that can fit on a standard plate. At a minimum, it requires the following:

a. Material: carbon steel or stainless steel
b. Threaded shaft
c. Slotted or Phillips head
d. Zinc or other surface coating
e. Energy (electricity)
f. Packaging material

This list is referred to as a Bill of Materials (BOM). Irrespective of whether a product is hardware or software, it has a BOM. For instance, the number of components that go into an iPhone 3G can be estimated to be at least a hundred,[51] and the count in the latest models is significantly higher. Even purely software products publish a Software BOM (SBOM), detailing the various modules they incorporate.

The BOM and its cousin SBOM are recursive in nature. Each component in the list may not be a final product in itself and may have its own BOM or SBOM. When fully expanded, the complete BOM can run into thousands of components, forming a supply chain. When represented mathematically, this chain resembles a graph. Since not every component can be produced by the same vendor, manufacturing a product requires managing the inputs and outputs across this network.

When most components are available within a specific geographical area, manufacturers save on logistics costs. As a result, the more complex and extensive a product's BOM, the more production tends to cluster geographically. Over time, this leads to the emergence of industrial clusters—regions where the entire supply chain for a product is co-located. These clusters offer several advantages: shorter lead times, easier collaboration, lower transportation costs and faster innovation through tighter feedback loops. A classic example is Shenzhen in China, which hosts a dense ecosystem of suppliers across printed circuit boards, casings, assembly plants, testing services, packaging, logistics and even chip manufacturing.

Software, however, behaves differently from hardware manufacturing. It does not require geographically concentrated physical clusters for several reasons:

1. Its 'bill of materials' consists of code packages rather than physical components.
2. Distribution occurs through platforms such as npm, GitHub and Docker Hub, rather than physical logistics (though interestingly, Docker itself is modelled as a software 'shipping container').
3. The rise of open-source software allows developers to import libraries—whether for user interfaces, databases or machine learning—from anywhere in the world instantly and at negligible cost, making physical proximity irrelevant.
4. Cloud providers (AWS, Azure, GCP) eliminate the need for local server infrastructure; development, testing,

deployment and monitoring can all be handled remotely.

5. The asynchronous nature of software development enables global collaboration (for example, through GitHub pull requests).

LLMs, however, are neither purely software nor purely hardware. They are hybrid systems, emerging from the intersection of software, statistical modelling and intellectual property, running on specialized infrastructure. If one were to apply a BOM framework to LLMs, it would look as follows:

Level	*Layer*	*Components/Subsystems*
0	Final Product	• Deployed LLM (e.g., MyLLM-7B-v1.0) • Inference API/chatbot/embedded runtime/containerized package
1	Major Assemblies	• Model weights and configuration • Fine-tuning data • Deployment wrapper (API, CLI, edge runtime)
2	Subsystem Components	• Transformer architecture (layers, heads, activations) • Training framework
3	Foundational Materials	• Raw datasets (Common Crawl, books, code) • Software dependencies • Machine learning libraries
4	Optimized Cloud Infrastructure	• Custom chips and optimization libraries
5	Physical Infrastructure	• Data centres • Electricity and cooling infrastructure

If each of these layers is treated as part of a supply-chain cluster and mapped geographically, the distribution appears as follows:

Level	*Layer*	*Geographical clusters*
0	Final Product	Global, but concentrated in the EU, China and the US
1	Major Assemblies	Global open-source contributors; large-scale training concentrated in the US (OpenAI, Anthropic) and China (Baidu, Tencent, Alibaba/ Qwen), with code and models also from the EU (France)
2	Subsystem Components	Frameworks and tooling led by the US (PyTorch by Meta AI), France/US (Hugging Face), and Germany (LAION datasets)
3	Foundational Materials	Data sources are global (Common Crawl, GitHub); OSS development concentrated in the US, China, Germany, India and the UK
4	Optimized Cloud Infrastructure	Highly concentrated in the US and China; GPUs (NVIDIA A100/H100) are manufactured in Taiwan (TSMC) and controlled by US-based NVIDIA; cloud infrastructures are mostly in the US; China has large state-owned GPU clusters (39 data centres with ~115,000 H100) despite export bans
5	Physical Infrastructure	Global, but the cost of electricity and cooling is much lower in some geographies

These supply-chain clusters (US, China, EU) map closely to the regions where foundation models have emerged. One notable exception is the UAE, which has prioritized model development (Layer 1) while aggressively investing in compute and physical infrastructure (Layers 4 and 5).

This analysis suggests the following hypotheses regarding the emergence of foundation models:

1. Regions that dominate the LLM supply chain are also the regions from which foundational models originate.
2. Regions that control multiple layers—especially compute (Levels 4–5), model development (Levels 1–2), and core

software infrastructure (Levels 2–3)—are most likely to produce foundation models.

3. Without control over at least three of these layers, a region is likely limited to fine-tuning or hosting existing models, rather than originating them.

The Transformer Model and Its Dependence on GPUs

LLMs are built on the transformer model,* which is based on multiple attention layers and multiple attention heads within each layer. To understand this, consider the problem of monitoring activity in a 10-floor building using surveillance cameras. One would naturally ensure that each floor has cameras covering all areas, with overlapping angles to maximize visibility. If we standardize 10 cameras per floor, each floor becomes a *layer*, and each camera represents a *head* within that layer.

Now imagine a hypothetical human being with 100 heads and eyes, all connected to a single brain. This being can view all the camera feeds simultaneously, analyse them individually and also synthesize them into a unified understanding over time. Such a being would be able to form a high-level understanding of what is happening across the entire building.

This hypothetical observer does not need to store all the video footage. Instead, it extracts patterns within each floor by paying attention—for example, to how people move and whom they interact with. It then creates abstract representations of these patterns at the floor level and combines them into a holistic view of the entire building. In effect, it builds abstractions on top of abstractions.

By simultaneously attending to multiple cameras and multiple aspects of the footage, this system continuously transforms raw data into insights. It could, for instance, detect if someone on the

*Vaswani, Ashish, Noam Shazeer, Niki Parmar, Jakob Uszkoreit, Llion Jones, Aidan N. Gomez, Lukasz Kaiser, and Illia Polosukhin, 'Attention Is All You Need', 2023, https://arxiv.org/abs/1706.03762. Accessed on 25 May 2026.

seventh floor is behaving suspiciously. Now imagine replacing this hypothetical human with a computer system. What would it require?

- First, continuous access to large volumes of data (the camera feeds).
- Second, many parallel processors, each focusing on different aspects of the data.
- And finally, the ability for these processors to communicate with each other rapidly to form abstract representations of not just individual floors but also of the entire building.

In computing, two types of processors are relevant in this context: CPUs (e.g., x86 and ARM) and GPUs (e.g., NVIDIA, AMD Radeon). CPUs are designed for sequential processing and complex calculations, while GPUs are designed to accelerate image rendering and animations for video games and 3D applications. Almost all these tasks involved handling large matrices and performing calculations on them.

For instance, rendering a scene in a video game requires drawing millions of pixels continuously on a screen, all of which can be computed independently. A single GPU is hence made up of hundreds or thousands of smaller processing units—which can't do all complex calculations that a CPU can—that excel at performing simple operations on large datasets, typically represented as matrices.

Transformer models rely on deep neural networks that repeatedly perform such matrix operations—addition, multiplication and transformation—at scale. Core techniques such as gradient descent, forward passes and backpropagation are fundamentally applications of linear algebra, albeit at a massive scale. Recognizing this alignment, researchers around 2010 began repurposing GPUs—originally designed for graphics—to train neural networks.

A key breakthrough came in 2012, when AlexNet was trained on just two commodity GPUs and dramatically outperformed competitors in the ImageNet Large-Scale Visual Recognition

Challenge. This marked a turning point, demonstrating that deep learning was not merely theoretical but practically powerful.

Following this, researchers across domains—computer vision, speech recognition and natural language processing—rapidly adopted deep neural networks. Technology companies such as Google, Facebook and Microsoft scaled these approaches, driving massive investment in GPU-based AI infrastructure. NVIDIA, in particular, pivoted towards AI and became one of the world's most valuable companies by 2024.

The interdependence between GPUs and LLMs has deepened over time:

a. **Pascal (2016):** Enabled multi-GPU communication for large-scale training.
b. **Volta (2017):** Introduced tensor cores for efficient matrix operations.
c. **Ampere (2020):** Optimised for training large models.
d. **Hopper (2022):** Designed specifically for transformer workloads, with improved interconnects.
e. **Blackwell (2024–25):** Focused on massive memory bandwidth and large-scale inference.

The evolution of GPUs has closely tracked the development of more powerful AI models. For example, GPT-1 (2018), GPT-2 (2019), and GPT-3 (2020) were trained on Volta architectures; GPT-3.5 (2022) likely used Ampere; GPT-4 (2023) used a combination of Ampere and Hopper; and later models such as GPT-4.5 and GPT-5 are associated with Hopper and Blackwell systems.

A clarification is necessary here. While LLMs depend on transformer architectures and transformers benefit from GPUs, the hardware itself need not be from NVIDIA. Any system capable of large-scale parallel matrix computation with sufficient memory bandwidth could, in principle, be used—though NVIDIA currently leads in performance and ecosystem maturity.

In the LLM landscape, GPUs and data centres constitute critical physical infrastructure and are highly capital-intensive. The scale of investment required is immense. A McKinsey report[52] estimates that by 2030, around $5.2 trillion may be needed to meet global AI compute demand. A 2025 analysis of 500 AI supercomputers suggests that xAI's Colossus system alone costs approximately $7 billion in hardware and consumes around 300 MW of power—comparable to a small city.[53] The same analysis projects that by 2030, a single AI supercomputer could require $200 billion in hardware and up to 9 GW of power, equivalent to the output of nine nuclear reactors. It further points out the high geographical concentration of these AI supercomputers: approximately 75 per cent are in the US and around 15 per cent in China.

This dominance of clusters is a major reason why these regions have produced foundational LLMs. However, it is not a strictly necessary condition, as the example of France suggests.

Mistral: A Research Power

Mistral was founded in April 2023 by three AI researchers:*

1. Arthur Mensch (CEO), previously at Google DeepMind
2. Guillaume Lample (Chief Scientist), previously at Meta AI
3. Timothée Lacroix (CTO), also previously at Meta AI

They had a distinct vision: that open-source AI would outperform closed systems (e.g., OpenAI, Perplexity) and was essential for democracies. They prioritized architectural innovation, transparency and sovereignty. Their backgrounds, and their recognition as top-tier AI talent, led to a rapid influx of capital into the company.

In June 2023, just four weeks after founding, Mistral AI raised[54] a €105 million seed round at a valuation of €240 million, even

*'Frontier AI. For All of Us', *Mistral AI*, https://mistral.ai/about. Accessed on 25 May 2026.

before building a product or releasing a model. Shortly thereafter, in December 2023, it raised another €385 million at a valuation of $2 billion. By June 2024, a combined €600 million round had increased the company's valuation to €5.8 billion (~$6.2 billion).[55] Their first model, Mistral-7B, was released in September 2023[56]—just four months after the initial fundraise—and, as promised, incorporated architectural innovations that allowed it to outperform larger models such as LLaMA-13B.*

Public comments by Mensch illustrate the centrality of talent (quote lightly edited for clarity):[57]

> We saw a community building on top of LLaMA at the time, on top of LLaMa 7B. But LLaMa 7B wasn't good enough. And so, we realized that we could make it much better. We could make a much better 7B model. And so that's the sweet spot we targeted for our introduction to the world. And basically, we had to build the entire stack from scratch. So, getting the data, building the training code, getting the compute, which was a bit of a challenge because in these four months we were ramping up. So, we started at zero GPUs, and we **trained on like** 500 GPUs, 7B, I guess we went fast because the team was very motivated and therefore did not have a lot of holidays during these four months. And AI teams that succeed and **go** are typically **like** four to five people. **And AI teams that invent things have always been this size.** So, we are trying to have an organization where we have squads of five people working on data, working on pretraining, and so far, this has worked out quite well.

*The number of parameters in a model indicates its complexity and capability. In general, models with a larger number of parameters tend to perform better than models with fewer parameters. Therefore, when Mistral-7B outperformed the LLaMA-13B model, it was not merely surprising; it demonstrated that architectural innovations in transformers can matter far more than the sheer availability of compute.

The interview highlights how strong research talent enabled rapid hiring through networks at DeepMind and Meta, accelerated capital raising, and allowed the team to build and release a model within four months. It also enabled access to larger computing clusters emerging across geographies (e.g., CoreWeave's scalable GPU cloud).

Mistral also aligns with the emerging concept of AI sovereignty in the European Union—the ability of the region to develop, control and deploy AI systems without reliance on foreign (particularly US or Chinese) platforms; to ensure alignment with local values, laws and norms (e.g., GDPR, the AI Act); and to maintain control over critical infrastructure such as GPUs, cloud compute and large-scale models.

Mistral thus represents a textbook example of how strength in one layer of the AI supply chain—often implicit, such as human talent—can enable movement into adjacent layers (model development and architecture), and subsequently into others (compute infrastructure), supported by capital. Embedding within any one part of the supply chain can generate momentum across others, but only when guided by a deliberate strategy backed by both talent and capital.

Three-Power Model and LLM Supply Chain

There are three forms of power in the modern world: industrial power, financial power and research power. LLM foundation models tend to emerge only from locations—nation states or blocs—that possess at least one of these forms at scale.

Consider the US. While it no longer dominates semiconductor manufacturing, it remains a primary locus for foundation model development.

An industrial power can be understood as having the capacity to produce or control critical physical infrastructure (e.g., chips, compute clusters). The US, despite offshoring fabrication, retains:

1. **Design leadership:** Companies such as NVIDIA, AMD, Intel and Apple design the world's most advanced chips.

2. **Control over the industrial stack:** The US dominates electronic design automation (EDA) tools (e.g., Synopsys, Cadence), key intellectual property layers and strategic choke points in the semiconductor supply chain through export controls and alliances.
3. **Cloud infrastructure:** Amazon, Google and Microsoft collectively control a significant share of global data centre capacity, representing a new form of industrial power.

Financial power enables the mobilization of large-scale capital for R&D, compute infrastructure and talent acquisition. The US venture capital ecosystem is the most risk-tolerant globally, funding high-risk AI ventures such as OpenAI, Anthropic and Cohere. Large technology firms (Google, Microsoft, Amazon, Meta) possess cash reserves in the hundreds of billions, allowing them to fund multi-billion-dollar training runs and shape entire ecosystems through partnerships (e.g., Microsoft's $13 billion investment in OpenAI). Capital markets further reinforce this by rewarding perceived AI leadership and penalizing hesitation.

Research power refers to the concentration of elite institutions, talent pipelines and innovation ecosystems. The US hosts leading AI research universities (e.g., MIT, Stanford, Berkeley), attracts global talent (historically through immigration pathways), and integrates public and private research funding (e.g., NSF, DARPA).

This highlights a critical point: LLM foundation models are as much intellectual, financial and societal constructs as they are industrial ones.

China Presents a Contrasting but Equally Powerful Configuration

Industrially, China is the global hub for manufacturing and infrastructure build-out. It produces a substantial portion of the world's electronics and has deep capabilities in assembling servers, building data centres and operating cloud infrastructure at scale.

Companies such as Alibaba, Tencent and Huawei operate hyperscale infrastructure within China's digital ecosystem, supported by relatively abundant energy and cooling capacity.

Financially, China has made AI a national priority, directing large-scale public investment into infrastructure, semiconductor R&D and AI startups. Major technology firms (Baidu, Alibaba, Tencent, ByteDance) possess substantial balance sheets and actively fund LLM development. While its financial system is less venture-driven than that of the US, it is highly effective at allocating capital towards strategic objectives through industrial policy and state-aligned investment mechanisms.

In terms of research power, China has a vast and growing pool of technical talent, producing more engineers and computer scientists annually than any other country. Its research output—in terms of papers and patents—is substantial. Institutions such as Tsinghua University and Peking University, along with companies like Baidu and Huawei, are advancing LLMs tailored to Chinese linguistic and cultural contexts.

A striking example of convergence across all three dimensions is DeepSeek. Originating as a project within a hedge fund—entities designed to pursue asymmetric returns through high-risk strategies—its approach aligns well with foundation model development, where high upfront costs can yield foundational infrastructure for downstream applications.

Quantitative hedge funds attract elite talent in mathematics, statistics and programming—skills closely aligned with deep learning. Constrained by US export restrictions on advanced GPUs, Chinese AI labs, including DeepSeek, were forced to operate with limited access to cutting-edge hardware. To compensate, DeepSeek employed low-level optimizations such as PTX[58] (closer to assembly-level programming than high-level abstractions), along with other efficiency techniques, to achieve high model performance at significantly lower cost—surprising the US-led AI ecosystem.

Weaponizing Inter-Dependence and Escape Velocity

The LLM supply-chain perspective also reveals emerging geopolitical power dynamics, increasingly shaped as a contest between nation-state blocs. As described in earlier chapters, until the early 2000s, the US was both a leading industrial power and a financial power, and consequently a research power. However, accelerating financialization led to de-industrialization. In parallel, the adoption of a sanctions regime—particularly restricting access to high-end GPUs for China—triggered a search for alternatives.

These alternatives have taken multiple forms. Mistral demonstrated that architectural innovation—within the broadly known transformer paradigm—can occur even in resource-constrained environments, where export controls have limited relevance. DeepSeek, by contrast, showed that constraints can lead to novel optimization pathways, uncovering techniques that were rare even within NVIDIA and largely unexplored by other research labs.

Huawei has pursued a different path, developing software adapters that allow existing AI applications—built on widely used frameworks such as PyTorch[59]—to run on its own chips (e.g., Ascend NPUs). It is also building independent software frameworks[60] to reduce reliance on NVIDIA's ecosystem. These chips and frameworks currently achieve roughly 60 per cent of the performance[61] of the NVIDIA GPUs accessible to China, with expectations of narrowing this gap over time. In parallel, Huawei is investing in interoperability layers that allow models trained on NVIDIA GPU[62] to run seamlessly on alternative architectures, including its own.

The central premise underlying these approaches is the reduction of both training and inference costs. In the LLM world, the cost of training involves building models by updating parameters based on data and the cost of inference involves using trained models to generate output (e.g., answering questions, summarizing text). The goal is to drive inference costs towards near zero and minimize training costs as far as possible.

Ben Thompson captures how this strategy aligns with China's broader industrial approach:[63]

> This makes sense beyond idealism. If models are commodities—and they are certainly looking that way—then long-term differentiation comes from having a superior cost structure; that is exactly what DeepSeek has delivered, which itself is resonant of how China has come to dominate other industries. This is also contrary to how most US companies think about differentiation, which is through having differentiated products that can sustain larger margins.

In other words, Chinese firms are treating AI models much like manufactured goods such as screws or fasteners: mass-producing them on domestically controlled hardware and software stacks, investing in compatibility layers that allow deployment across different systems, and exporting them at extremely low cost. This reflects the logic of an industrial power—one that seeks not only to counter a sanctions-driven financial model, but also to establish an alternative development (referred to as social industrialism in the earlier chapter) paradigm.

The trajectory so far suggests that the US strategy of weaponizing interdependence through sanctions has not achieved its intended outcomes.

NVIDIA's CEO acknowledged this at the Computex in Taipei:*

> The local companies are very, very talented and very determined, and the export control gave them the spirit, the energy and the government support to accelerate their development.

---

*Davidson, Helen, 'US Chip Export Controls Are a "Failure" Because They Spur Chinese Development, Nvidia Boss Says', *The Guardian*, 21 May 2025, https://www.theguardian.com/technology/2025/may/21/us-chip-export-controls-a-failure-spur-chinese-development-nvidia-boss-says. Accessed on 25 May 2026.

He also added:

> I think, all in all, the export control was a failure. China has a vibrant technology ecosystem, and it's very important to realize that China has 50 per cent of the world's AI researchers, and China is incredibly good at software.

In effect, by weaponizing interdependence, the US may have inadvertently enabled China to develop escape velocity.

LLM Foundation Models Can't Emerge unless Structural Issues Are Addressed

The LLM supply chain can be mapped to the three-power model as follows:

Level	*Layer*	*Power mapping*
0	Final Product	Financial power
1	Major Assemblies	Research power + Financial power
2	Subsystem Components	Research power + Financial power
3	Foundational Materials	Research power
4	Optimized Cloud Infrastructure	Industrial power
5	Physical Infrastructure	Industrial power

If we map India onto this framework to try to understand the gaps, a slightly depressing picture emerges:

1. **Level 0:** India primarily acts as a consumer of global models (e.g., GPT-4, Claude, Gemini). Domestic startups largely focus on fine-tuning or building smaller models (e.g., Sarvam AI,*

*An LLM is considered 'large' if, on average, it is a ~7B parameter model, and it is considered foundational if it is not based on fine-tuning other LLMs. As of the writing of this book, Sarvam has released 3 models—Sarvam 1, 2 and M. 1 and 2 are small

Krutrim,* BharatGPT†).

2. **Level 1:** India does not host global-scale model training or original architecture design; contributions are limited to incremental tuning and dataset creation.
3. **Level 2:** No major toolchains, frameworks or foundational architectures have originated in India (e.g., PyTorch, TensorFlow, Hugging Face are all external).
4. **Level 3:** India does possess high-quality labelled datasets for local languages (e.g., Naamapadam,[64] Vakyansh,[65] iNLTK[66]).
5. **Level 4:** India lacks unrestricted access to export-controlled NVIDIA hardware and operates under quota-based access. While some GPU clusters exist through public–private initiatives, there is no sovereign ownership of advanced chips, large-scale compute infrastructure or optimization toolchains.
6. **Level 5:** There is no coordinated national strategy linking geography, energy and infrastructure to build a competitive compute advantage (unlike China's western regions or the US's Pacific Northwest).

This reality, however, is not immutable. In the following chapters, we outline a series of interventions and strategies that could address these gaps and improve India's position across multiple layers of the supply chain.

---

language models with 2B parameters, and M is a 24B parameter model based on Mistral.

*Krutrim claims that its models are frontier and foundational models. However, serious doubts persist regarding these claims. One important indicator is the limited adoption of its LLMs and cloud products. As reported by Swathi Moorthy in *The Economic Times*, Gartner analysts argue that Indian AI models do not stand up to leading global models because they are not sufficiently open to external scrutiny. See Moorthy, Swathi, 'Krutrim Finds Few Takers for Its LLMs and Cloud Products', *The Economic Times*, 3 June 2025, https://economictimes.indiatimes.com/tech/artificial-intelligence/krutrims-ai-products-not-seeing-enough-takers-say-some-experts/articleshow/121576245.cms. Accessed on 25 May 2026.

†The README.md of the repository explicitly states that the model is fine-tuned. See *BharatGPT-3B-Indic*, Hugging Face, https://huggingface.co/CoRover/BharatGPT-3B-Indic/blob/main/README.md. Accessed on 25 May 2026.

5

How Chinese Scammers Built Up a Money Laundering Network on Top of UPI

The crime boss had a unique problem. It was not one caused by the failures of his illegal trade, as had been typical in his early years of business. His problem of late was success—his wild success. His land of birth, China, had a weak legal system dominated by the state and state-backed enterprises. Private businesses were harder to run and hence dependent on *guanxi*,* a path in which he saw neither independence nor sustained profitability. So he formed his own gang and, through guanxi, built his business over time. In the initial days, it was subsistence; but as it grew, he began generating more and more free cash that needed to be managed. For that, he had no choice but to lean on formal financial channels.

He found a neat way to manage the surplus: peer-to-peer (P2P) lending. His criminal network had already built the infrastructure required to collect any loans disbursed to borrowers (collateral and muscle). The near-absence of regulation in 2012 meant he could legitimately build a lending platform powered by technology. From

*Quite literally, personal connections that help businesses. For an in-depth treatment of guanxi, see Xin, Katherine K., and Jone L. Pearce, 'Guanxi: Connections as Substitutes for Formal Institutional Support', *Academy of Management Journal*, Vol. 39, No. 6, 1996, pp. 1641–1658.

his background in extortion, he knew that the best collateral was not physical assets but information assets. He often told hesitant business owners that their shops were not his collateral, but their spouses and kids—and this usually resulted in immediate repayment of their dues.

But scaling this model across the geographies he controlled required data. In his experience, everyone lived on their smartphones and did not really understand that granting access to data on those devices meant surrendering their entire lives. That borrowers would do so for a trivial reduction in interest rates always amused the criminal in him. 'More for nothing,' he would exclaim, displaying the nudes extracted from their phones to delinquent borrowers, who then promptly cleared their dues. The returns he generated soon became comparable to the money he made from his other businesses. As word spread, investors and ordinary citizens alike wanted a share of the action.

How P2P Lending Spread and Ended in China

The one thing the crime boss understood well was that criminals who specialize in *wet work* tend to die young. He was therefore always looking for ways to reduce his mortality risk while expanding his earnings. The P2P business opened his eyes to such an opportunity. All that was required was to take money from lenders (who are often not that smart), lend it to borrowers and take a small cut along the way. At scale, this could make him very rich, without the risks involved in wet work.

This model—often referred to as a *pure information intermediary*—worked well. In fact, it worked so well that it attracted competition. A pure information intermediary connects lenders to borrowers using data accumulated by the platform. This is particularly effective in segments where formal credit scores are absent. There are two primary reasons for this absence:

1. A large segment of the population earns too little, making them unattractive to traditional financial institutions given the cost of underwriting and collection.
2. Another segment earns enough but relies on informal credit networks, often mediated through guanxi.

The platform he built focused more on the second category than the first. While he had heard of complex algorithms used in the West to assess borrower risk, he relied on a simpler heuristic: raising income levels. The party's push for making China a manufacturing hub had a three-step playbook:

1. Financial repression*—a phenomenon whereby interest rates were artificially suppressed, allowing favoured borrowers (national champions) to access cheap capital.
2. Capital allocation—this cheap capital flowed into export-oriented industries, strategic sectors, large manufacturers and local governments, leading to deliberate overinvestment and excess capacity.
3. Export-led growth—overcapacity drove down global prices, making exports competitive and increasing income levels across the supply chain.

In the geography where he operated, these dynamics ensured rising incomes, which in turn reduced default risk. Whatever residual risk remained was managed through enforcement mechanisms and the data he collected. In quiet moments, the crime boss reflected that he had become a white-collar operator engaged in dubious practices—a significant evolution from his wet work criminal background.

---

*Tao, Yu, and Shen Wei, 'Funds Sharing Regulation in the Context of the Sharing Economy: Understanding the Logic of China's P2P Lending Regulation', *Computer Law & Security Review*, Vol. 35, No. 1, February 2019, pp. 42–58, https://doi.org/10.1016/j.clsr.2018.10.001. Accessed on 25 May 2026.

The information intermediary model works by positioning the platform at the centre of cash flows. Both lenders and borrowers create 'accounts' on the platform, providing identity, bank details and related information. When a borrower seeks a loan, they specify the amount, tenure and preferred interest rate within a given range. The platform assigns a 'rating' (a form of credit score) and presents the request to lenders. The lender deposits the amount with the platform, which deducts its cut and routes the remainder to the borrower. Borrowers also repay through the platform.

This central position in cash flows—without directly assuming credit risk—allows the platform to function somewhat like a credit bureau[67] by mining patterns of behaviour from both lenders and borrowers, but with potential for higher revenue.* However, unlike regulated credit bureaus, information intermediaries are exposed to competitive pressures on both supply (lenders) and demand (borrowers) side. For such a platform to scale, it must constantly expand both sides, as not focusing on one side and letting it down would have an immediate impact on the other.

Given that the pool of lenders and borrowers is finite within a geography, growth typically follows two paths:

- Geographic expansion, enabling access to new sets of borrowers and lenders and arbitrage opportunities.
- Fee compression, reducing commissions to attract higher transaction volumes.

Both these approaches always result in margin compression.

This problem also gets compounded when borrowers default during economic downturns. The assumption of continuously rising income levels across the borrower base is a mathematical impossibility given the economic ups and downs. As defaults

---

*Credit bureaus charge a fixed price for credit data, whereas P2P information intermediaries charge a fee that is a percentage of the loan amount. Hence, over a period of time, they are able to earn a higher income compared to credit bureaus.

increase, platforms lose both lenders and borrowers, forcing them to adopt new models to remain competitive. This typically leads to a single outcome: credit default guarantees, where the platform guarantees the borrower a cushion against borrower default.

When the platform assumes credit risk, it will always be fractional, and it can only be funded in a few ways:

- Deducting a 'guarantee fee' from the loan disbursed to the borrower.
- Retaining a portion of the lender-side commission as a reserve.
- Allocating a portion of the platform's equity to absorb losses.

Regardless of the mechanism, the outcome is the same: the platform effectively becomes a *shadow bank*—but without regulatory oversight.

The absence of oversight creates fertile ground for fraud. Several examples illustrate this:

- **Ezubao:*** Functioned largely as a Ponzi scheme; apart from one genuine borrower, most were fictitious.
- **Fanya Metals:†** Collected investor funds through an unregulated P2P structure before halting payments.
- **Shanlin Finance:‡** Another large-scale Ponzi operation.

Sunny Huang applied a Nash equilibrium framework[68] to P2P lending and concluded that, in such environments, sustained

*Mitchell, Tom, 'Arrests in China over $7.6bn Ponzi Scheme', *Financial Times*, 1 February 2016, https://www.ft.com/content/4ca011f4-c88f-11e5-a8ef-ea66e967dd44. Accessed on 25 May 2026.

†Hornby, Lucy, 'Head of Fanya Metals Scheme Arrested in China', *Financial Times*, 23 June 2016, https://www.ft.com/content/10b02844-38df-11e6-9a05-82a9b15a8ee7. Accessed on 25 May 2026.

‡Tang, Frank, 'Eight Charged in China over "Ponzi Scheme Posing as P2P Lender" that Took US$9 Billion', *South China Morning Post*, 25 April 2018, https://www.scmp.com/news/china/economy/article/2143299/eight-charged-china-over-ponzi-scheme-posing-p2p-lender-took-us9. Accessed on 25 May 2026.

profits are often only achievable through fraud. This is because fraud is the only variable not dependent on the success probability of the underlying loans.

The slow rise of fraudulent platforms was driven by a combination of factors: financial repression, industrial policy that increased income levels but did not provide adequate avenues for formal returns and the complete absence of regulation in the P2P sector until 2015—even though the first P2P lending company, CreditEase (YiXin), was set up in 2006. The first set of regulations published in 2015[69] was largely advisory in nature and encouraged internet-based finance, adopting a hands-off approach.

In 2016, regulations* were introduced to reduce operational risk. These required the separation of platform-owned funds from client funds, with the latter to be held in third-party custodian accounts (preferably banks), and explicitly prohibited embezzlement. Within a short span of eight months, these regulations were further tightened. Caps were introduced on the amount that could be lent to individuals or entities, and P2P companies were required to register with both local provincial governments and telecom authorities. In effect, these measures began to formalize P2P platforms into quasi-credit bureaus, imposing due diligence requirements similar to those of banks, thereby making the pure information intermediary[70] model increasingly unviable.

In 2018, a further set of regulations prohibited the shadow banking model[71] (i.e., deposit guarantees) and made it clear that only a limited number of entities would be permitted to operate. These measures led to the rapid closure of P2P platforms: from 5,970 in 2017 to effectively zero by November 2020.[72] While it may be argued that regulators overreached, from their perspective,

*State Council of the People's Republic of China, General Office of the State Council, 'Notice on the Issuance of the Implementation Plan for the Special Remediation of Internet Financial Risks', No. 21, 4 April 2016, https://www.lawinfochina.com/Display.aspx?lib=law&Cgid=282015&EncodingName=big5. Accessed on 25 May 2026.

the opacity of these platforms and their business models made it impossible to distinguish legitimate operations from Ponzi schemes or unregulated shadow banks. Shutting down the sector became the only viable option.

The scale or prominence of these platforms did not offer protection. This was demonstrated when the IPO of Ant Financial was halted on the orders of Xi Jinping* on 3 November 2020, following a speech by its founder, Jack Ma, criticizing regulatory authorities. Ant Financial had stepped into the gap left by the collapse of P2P platforms. By leveraging extensive customer profiling—drawing on social media activity, communication patterns and spending behaviour—it was able to connect borrowers nationwide with provincial lenders.

The systemic risks posed by such platforms were subsequently addressed through the '30 per cent rule', which required financial intermediaries to fund at least 30 per cent of any loan disbursed outside the lender's licensed geography. Depending on its application, this effectively transformed platforms like Ant from intermediaries into banks.

How Chinese P2P Lending Morphed into Chinese Loan Apps in India

Chinese entrepreneurs derived several key insights from their experience with internet finance via their P2P ventures:

- Finance is one business where the raw material and the final product are the same—money—and demand is effectively infinite. What matters, therefore, is distribution (capturing demand) and collection. In both cases, technology can make a difference—the more coercive and underhanded

*Yang, Jing, and Lingling Wei, 'China's President Xi Jinping Personally Scuttled Jack Ma's Ant IPO', *The Wall Street Journal*, 12 November 2020, https://www.wsj.com/articles/china-president-xi-jinping-halted-jack-ma-ant-ipo-11605203556. Accessed on 25 May 2026.

the technology, the better it is for addressing demand capture and collection. Hence, 'internet finance' is not two separate domains but a fused system, where technology and finance are one and the same. This fusion means regulatory controls can be easily overridden by exploiting knowledge asymmetry.

- Regulators typically govern industries by defining permissible practices and licensing entities accordingly. By nature, they assume that intended practices are expected practices and often conduct audits on entities to understand gaps and perform remedial measures. This works in stable, predictable domains. However, in finance—where demand is effectively unbounded—new practices can emerge outside the regulatory imagination and rapidly capture market share through unconventional or 'guerrilla' tactics.
- These tactics exploit knowledge asymmetry: regulators often have an incomplete or imprecise understanding of technological developments, leading to an underestimation of risks and an overestimation of benefits. Given limited regulatory capacity, oversight is structured hierarchically, with greater scrutiny at the top and minimal attention at the bottom.
- In a domain with infinite demand, this creates blind spots. Shadow ecosystems emerge to serve unmet demand. And when this shadow industry organizes itself into entities that are also at the bottom tier of the pyramid, they create a perfect opportunity for takeover—first by influencing their technology stack and then, once enough insights are gained, integrating into the financial supply chain.
- Given the low frequency of scrutiny at these lower tiers, combined with naïve borrowers, such entities can extract significantly higher returns on capital than regulated

players. As these practices spread, a bandwagon effect emerges, reshaping the ecosystem itself.

- This resembles an Overton window shift, not in the information domain but in finance. Because finance is driven by information flows, the integration with technology accelerates these shifts dramatically.
- Once practices shift due to technological change, capital flows towards those with a technological advantage. In finance—where capital is both input and output—technological advantage ultimately dominates financial advantage.

In India, the lending landscape comprises three broad segments: banks, NBFCs and informal lenders. Banks dominate the formal lending ecosystem, benefiting from access to low-cost capital through savings and current accounts, as well as fixed deposits. Consequently, they tend to lend primarily to creditworthy borrowers.

NBFCs compete with banks for these borrowers but also serve segments that banks often avoid, due to factors such as limited branch presence or higher perceived risk. As a result, NBFCs are typically less risk-averse than banks. However, they also have more limited access to capital and therefore seek solutions that are cost-effective and easy to deploy. This often leads them to integrate existing technological solutions rather than build their own.

Chinese P2P players around 2016 had mature software stacks that handled all aspects of lending, including facial recognition, identity verification and digital distribution (via apps). While these stacks were initially built and owned in-house, as the market expanded rapidly after 2006, many components became available off-the-shelf, with integrations tailored for specific data sources in South-East Asia (including India).* For instance, Advance.AI offers

*See the Internet Archive Wayback Machine entry for Advance.AI's 'ID Document

both a technology stack addressing key digital lending use cases (KYC, fraud checks) and a digital finance arm (Atome Financial).[73]

The SDKs (software development kits) were simple to use and integrate, making them particularly attractive to NBFCs operating at the lower end of the market. These entities neither had the capital nor access to the talent required to build and maintain such technology stacks; their core strength lay in regional presence and brick-and-mortar operations. The promise these SDKs offered mirrored what P2P platforms had earlier provided to regional banks in China: connecting nationally distributed borrowers with capital providers who lacked the capability to build and maintain technology.

In many cases, merely providing SDKs was insufficient. NBFCs often lacked the capacity to keep pace with rapidly evolving mobile operating systems and web browsers. This led to outsourcing the development and maintenance of apps to small firms closely associated with SDK providers. As this pattern replicated across multiple NBFCs, it gave rise to more sophisticated tooling—platforms that could repackage the same underlying product with configurable interfaces, branding and logos, without requiring any change to the underlying code.[74]

The lending business itself can be modelled in two ways:

- As a set of interconnected supply chains
- As a layered stack imposed on those supply chains

First, consider the supply chain view from both financial and technological perspectives.

The financial supply chain is as follows:

Recognition', which specifically notes support for document recognition in India: https://web.archive.org/web/20201201212130/https://www.advance.ai/en/id-document-recognition. Accessed on 25 May 2026.

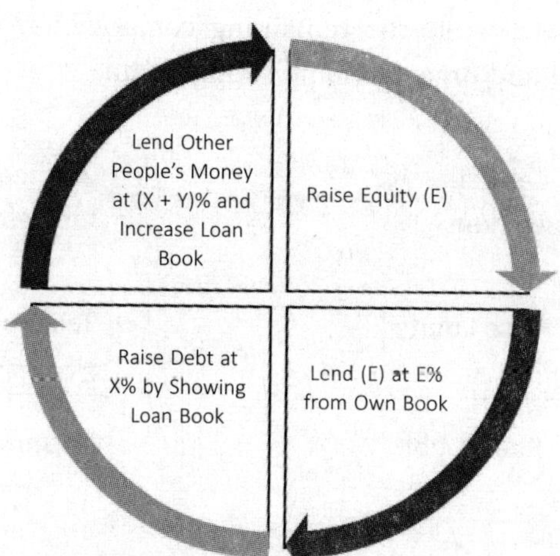

The technological and organizational supply chain required to enable the financial chain is as follows:

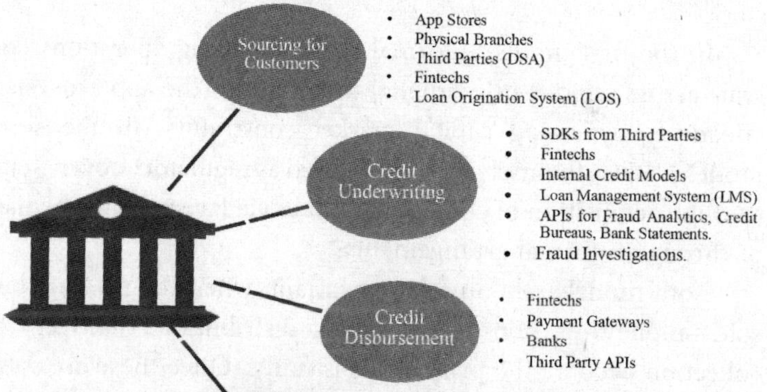

Another way to understand the relationship between the two is that the stages of 'raising equity' and 'raising debt' belong to the capital markets, while the remaining components constitute the technological and organizational supply chain.

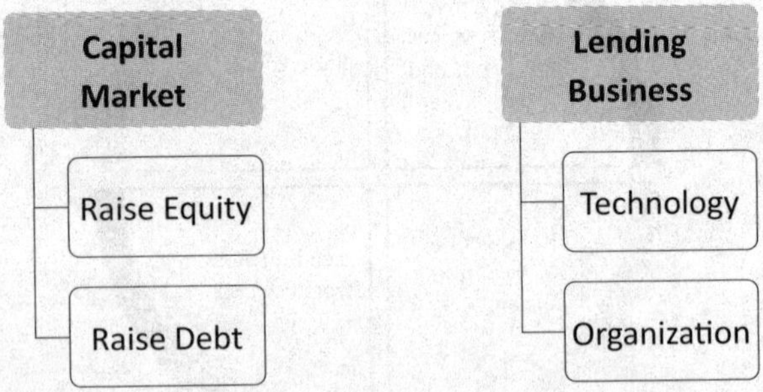

NBFC regulations in 2016 were primarily focused on the capital market side of this structure, while remaining relatively permissive on the lending side, with almost no oversight of the technology layer. This created two distinct pathways for moving up the lending stack.

In the first model, informal money-lending operations could scale across regions using digital distribution (the app ecosystem), effectively bypassing capital market constraints. In the second model, NBFCs themselves could be used as regulatory cover—either through acquisition of controlling stakes via layered intermediaries or through informal arrangements.*

Both models rely on a key invariant: when the product being sold—money—is non-differentiated, distribution, discovery and collection become the primary constraints. Once these are solved, capital market regulations can be effectively circumvented.

Distribution and discovery were addressed through app stores (both Android and iOS). Since NBFCs were already using third-party providers for distributing their lending apps, it became relatively easy

*This is the classic Benami approach, where the entity is controlled by third parties who have informal arrangement with the owners, by promising rewards for the cover, provided by their shareholding.

to create similar apps that were like that of the NBFCs or even smaller banks—by reusing or modifying existing supply chains.

In the initial phase, Chinese loan apps focused exclusively on app stores, leveraging their existing embedded position within NBFC technology ecosystems. With the rapid acceleration of digital adoption following the onset of COVID-19, demand for credit surged due to income disruption, while supply from both banks and NBFCs lagged significantly.

The table below summarizes the demand shock:*

Month-Year	*Rate*	*From*	*Difference*
April-2020	15.70 per cent	15.00 per cent	0.70 per cent
May-2020	10.60 per cent	16.90 per cent	-6.30 per cent
June-2020	10.50 per cent	16.60 per cent	-6.10 per cent
July-2020	11.20 per cent	17.00 per cent	-5.80 per cent
August-2020	10.60 per cent	15.60 per cent	-5.00 per cent
September-2020	9.20 per cent	16.60 per cent	-7.40 per cent
October-2020	9.30 per cent	17.20 per cent	-7.90 per cent
November-2020	10.00 per cent	16.40 per cent	-6.40 per cent
December-2020	9.50 per cent	15.90 per cent	-6.40 per cent
January-2021	9.10 per cent	16.90 per cent	-7.80 per cent
February-2021	9.60 per cent	17.00 per cent	-7.40 per cent
March-2021	10.20 per cent	15.00 per cent	-4.80 per cent

All that remained for the Chinese loan apps was to capture this demand by flooding app stores with instant-loan apps during the first phase. For borrowers during that period, facing extreme distress induced by COVID-19 lockdowns and the evisceration of

*It is obtained from the index page maintained by the RBI at https://rbi.org.in/Scripts/Data_Sectoral_Deployment.aspx. Individual monthly press releases were then examined to extract the year-on-year growth figures for personal loans for each month compared with the corresponding month in the previous year, and the data were subsequently collated into tabular form.

income, any lifeline delivered quickly was better than having none. Chinese loan apps were in a perfect position to capture this demand. Drawing on their experience in operating P2P lending stacks in their homeland, and having been booted out by Xi Jinping's regulatory actions, they found a ready and receptive market in India.

They already had technological mastery in running bot farms to artificially boost app store visibility and had the capital to run aggressive advertising campaigns. Users searching for 'loan' in app stores would typically see only these loan apps dominating the top results, supported by great reviews and ratings (created by bot farms). All of these apps promised disbursement within minutes and required minimal information—Aadhaar, PAN, selfies—and, of course, access to the entire phone (including contact lists, call records, SD drive, photos, videos, SMS messages, and even accessibility control that could install more software). The funds were disbursed instantly, as promised, which was good enough for borrowers in the initial phase.

The First Iteration of the Money Laundering Network

Irrespective of whether these Chinese loan apps had wink-and-nod arrangements with NBFCs or simply set up shell companies to disburse loans, there were two big problems that they had to solve:

1. How to collect EMIs and repayments from borrowers
2. How to repatriate the collected money back to China

The first problem was relatively easier to solve, given the rise of fintech platforms in India. Any digital business must support multi-channel payment collecting from their customers via net banking, credit cards, debit cards, UPI, wallets, auto-debits and mandates. Fintechs had already addressed this through the merchant account and nodal account model:

- A merchant (individual or business seeking to collect payments) creates an account with a fintech provider.

More often than not, all that was required was their GST registration, PAN, bank account number and, sometimes, KYC of their directors.

- Verification of GST and PAN is trivial through existing API providers (which the fintechs were using), while bank accounts could be verified instantly by the 'penny drop' protocol.*
- Once verified, the fintech assigns a nodal account, where incoming payments from customers are held separately from the fintech's own funds—aligned with regulatory requirements mandating segregation of customer funds.
- Fintechs provide APIs and payment interfaces (excluding cash), enabling customers to pay via multiple instruments. Funds are held temporarily in the nodal account, after which the fintech deducts its fee (Merchant Discount Rate, negotiated with the merchant) and transfers the balance to the merchant's registered bank account.

The combination of existing infrastructure and largely self-declared KYC processes made it easy to create multiple merchant accounts across fintech platforms. Chinese loan apps could therefore rapidly spin up new entities and route collections through multiple channels.

For instance, it is common practice for a single shell company to have multiple merchant accounts. The market dynamics of the merchant discount business meant competition was fierce, and the primary metric investors rewarded was GMV (Gross Merchant Value—money passing through their systems), which served as a proxy for revenue (a percentage of GMV). This meant fintechs prioritized onboarding more merchants, rather than scrutinizing

---

*Penny drop refers to a bank account verification method in which a very small amount—typically ₹1 along with a few random paise—is transferred to a bank account using the IMPS protocol. The receiving bank then confirms whether the account exists. Borrowers are often asked to provide the exact amount received, including the random paise component, as a way to verify access to the account.

what business those merchants were engaged in or how they generated income.

Features such as split payments, invoice links and QR codes—offered by fintechs[75]—became effective collection mechanisms for loan apps. With multiple shell companies[76] and multiple merchant accounts[77] per entity, collected funds could be layered extensively with minimal detection. The only requirement was technical sophistication, which these operators already had.

The second problem was how to launder (or convert the money earned from loan apps into legitimate income). Shell companies provided a convenient mechanism, as they could both earn income and incur expenses. For instance, assume a bank account opened by a fictitious person, who also opened a merchant account with one of the fintechs. When a loan app victim pays, the money first hits the nodal account and then the bank account opened by the fictitious person.

Once a certain amount accumulates, say ₹25,000, the controller of the fictitious account can withdraw the money via an ATM or AEPS, and re-deposit it as income in a cash-heavy business (e.g. petrol bunks, eateries, construction, wedding planning, small transportation companies), after taking a percentage (usually in the range of 1 to 2 per cent). Since cash is fungible, the money in these cash-heavy businesses effectively becomes white (laundered and hence treated as legitimate income).

In the next leg of the transaction, these businesses transfer a portion of this income to shell companies controlled by the operators (again after taking a percentage). In the final leg, the shell company, against this income, buys services[78] from another firm in either Hong Kong or Singapore and offsets the expense against the income.

But where does the money that the loan apps were disbursing come from? Loans must be funded by capital inflows from elsewhere, and while NBFCs and their directors might participate in benami arrangements—allowing loan operators to ride on their licences for a fee—they may not have the capital to lend. Injecting capital through acquisition is also problematic, as it is rigorously scrutinized by regulators.

It is widely known within export–import industries that under-invoicing of imports happens on certain products. For instance, if a product is worth $100 in reality, both the importer and the exporter collude to declare it as $50 to save on customs, GST and other taxes. But the importer still pays the exporter the remaining $50, and this has the same effect as illicit fund flows within the importing country, because of the following sequence:

A. When an importer brings in $100 and declares the full amount, there is no gap between legal and actual inflow.
B. However, if they bring in $100 but declare only $50, a gap emerges between declared and actual inflow. This difference is the illicit inflow.
C. A similar effect occurs when, instead of cash, the importer brings in goods or purchases services.

Now consider how two colluding parties on the exporting side can work together to fund the capital for the fake loan apps through the import under-invoicing model:

1. Say an exporter on the Chinese side under-invoices their export by $50.
2. They then have an arrangement with a local crime boss—previously in the P2P business that was shut down—to pay the remaining $50 in cash at a discount (say 10 per cent).
3. The crime boss can afford to do this because he is flush with cash and is looking for better investment opportunities.
4. The exporter benefits because this effectively acts as invoice discounting, and irrespective of the markup, the $45 earned is free money, as it is not declared as income.

But how does the crime boss recover the $50 (and earn a return), given that he has already disbursed $45 to the exporter? For that, he needs just one colluder on the Indian side. This colluder collects identity documents and creates a network of shell companies. The importer then purchases services from these shell companies and transfers the $50 owed to the exporter in Indian currency, at a slightly higher rate than the official exchange rate.

The shell company now has capital to lend, either directly or routed through NBFCs.

Beggaring Supply Chains and People

Recall the earlier discussion in Chapter 2 on how China dominates even the production of modern goods, and consider a recent conversation with a venture capitalist (VC). The VC had invested in a consumer goods company that initially performed well but had hit a roadblock: competitors were flooding the market with cheap imports from China, making it difficult to compete on price. The VC believed that imposing higher tariffs would help. His argument was as follows:

- The investee operates largely within the Indian market, with local supply chains and inputs.
- A minimum margin is required on unit cost, so prices are set accordingly.
- If the unit cost is ₹90 and the margin is 10 per cent, the selling price becomes ₹100. With 18 per cent GST, the market price becomes ₹118.
- However, competitors—mostly traders with no domestic manufacturing—sell the same product at ₹70, with a market price of ₹82, even though tariffs are 25 per cent.
- Even assuming Chinese production costs at ₹70, tariffs would raise this to ₹87.5, and with freight and other costs, it should match domestic pricing. Traders would also need at least a 10 per cent margin.

- Hence, the only viable solution seems to be higher tariffs.

This argument has several flaws:

1. A Chinese producer may accept a 10 per cent price reduction for higher volumes, bringing the cost down from ₹70 to around ₹63. Even with margin compression, total income may rise.
2. If the Chinese or local government treats the sector as strategic, it may subsidize freight, insurance and other costs—say by ₹3 per unit.
3. Trade-based money laundering then comes into play, where the item is under-invoiced by 50 per cent and shown as ₹30.
4. The effective unit cost to the importer becomes ₹30 (actual value) + ₹30 (paid to the shell company) + ₹7.5 (tariff), totalling ₹67.5.
5. Hence, if they can sell it at a market price of ₹70, they still make a tiny profit as the volume makes up for it, along with the fact that they have not invested anything.
6. Additional 'participation fees' may further distort pricing, where shell entities recycle funds by purchasing goods from the same trading networks.

Illicit inflows—via under-invoicing of imports—thus ended up funding an illicit credit boom through loan apps. These apps, in turn, extracted income from a population already weakened by the economic shock of COVID-19. That income was then channelled out of India through a mix of cryptocurrency trades and payments for non-existent digital services (via over-invoicing of imports), with everyone involved in the supply chain taking a cut.

A deficit of industrial inclusion, compounded by structural disabilities in India, meant that the VC's investee already faced a unit cost disadvantage. Government policy attempted to address this through tariffs. However, this was overridden by local collusion, enabled by weak regulatory oversight and limited enforcement capacity.

This local collaboration then enabled illicit capital inflows and outflows that not only undermined firms like the VC's investee but also extracted wealth from ordinary citizens.

And this 'beggaring' was enabled by digital technologies that the crime boss and his network had mastered through their experience in P2P lending. The digital–industrial–financial supply chains can thus be weaponized by private actors—and sometimes even shaped by state policies—to produce outcomes far more debilitating than conventional conflict. Yet such devastation rarely provokes equivalent outrage, because its effects are diffuse and not viewed holistically.

As Sunny Huang rightly pointed out before in the context of P2P lending, Nash equilibrium governs not just the returns of fraudulent enterprises but also the losses of nation-states. Ultimately, it all boils down to regulatory capacity.

The Indian (Regulatory) State Strikes Back (Imperfectly)

The persistent use of app stores for distribution, the takeover of NBFCs for legitimacy cover and the use of fintechs for collections prompted a series of regulatory actions. As is typical, the response took the form of licensing and disclosures, coupled with some truly bizarre moves that reveal the fragmented nature of the state's regulatory capacity.

First came regulations targeting app stores. They were expected to demand documentation from app developers proving that they were registered with the relevant authorities (e.g., NBFCs with RBI, investment platforms with SEBI, insurance with IRDAI). Restrictions were also imposed on loan tenure (e.g., very short tenures, such as one week, could lead to delisting), data collection practices (e.g., limiting access to contact information, call logs) and device identification (e.g., IMEI data being masked with pseudo-identifiers at the platform level).

This was followed by mandatory licensing for Payment Aggregators and Payment Gateways. All fintechs were required to obtain licences and were barred from onboarding new customers until approval was granted. Since denial of a licence would effectively shut down their business, this forced them to cooperate with investigations and tighten internal processes.

Subsequently, NBFC licences of entities involved in the supply chain were cancelled (e.g., X10 Financial Services,[79] Zavron Finance,[80] UMB Securities, Anashri Finvest, Chadha Finance, Alexcy Tracon, Jhuria Financial Services),[81] and new rules were introduced to regulate digital lending apps. NBFCs were also instructed to provide a list of apps designed and hosted in their name, enabling the regulator to publish a whitelist for app stores and users.

However, the URL published by the RBI was not search-friendly and even for us, it took 30 minutes of data analysis to understand whether a loan app is legitimate or otherwise. Further, several app links in the URL had no Play Store or App Store links and instead referred to only a web URL, which suggests that the list was self-reported and was not curated by the RBI. Fake loan apps continue to feature consistently in the top 10 and top 100 rankings, as extensively documented for over three years by researcher Babu Pooniawallah.* In our discussion with him, he pointed out that app stores often act only after sustained public pressure and naming-and-shaming—and even then, action is inconsistent. This is partly because they are global Big Tech firms headquartered in the West, with limited incentive to prioritize the laws and regulations of another state.

*Parasnis, Sharveya, 'Playing Cat and Mouse with Scammers: How Babu Pooniawalla Took Down over 8300 Fake Apps', *MediaNama*, 14 February 2025, https://www.medianama.com/2025/02/223-interview-babu-pooniawalla-taking-down-8300-fake-apps-google-apple-app-stores. Accessed on 25 May 2026.

How to Design an Illegal Payment Gateway

On 24 October 2024, the Ministry of Home Affairs issued a press notice* warning citizens not to rent or sell their bank accounts, Aadhaar or Udyam (small business) certificates to cybercriminals. The note stated that these were used to create illegal payment gateways, but did not explain how.

Consider the impossibility of the claim: fintechs that built legitimate payment gateways (e.g., RazorPay, CashFree, Paytm) raised billions in capital, yet cybercriminal networks—without such funding—have effectively built functioning payment gateways that evade law enforcement!

To understand how the scammers achieved this impossible feat, it is essential to recognize their motivation. The new Payment Aggregator and Gateway regulations prevented them from using existing payment gateways to collect money via the fake loan apps. They therefore had to invent a new method that functioned as a near drop-in replacement.

The essential components of a legal payment aggregator are:

1. Ledgers that track inflow of money into various merchant accounts.
2. A nodal account to hold incoming funds separately from the aggregator's own money.
3. Paying mechanisms to transfer funds to merchants after deducting fees and commissions.
4. Integration with APIs to track inflow and outflow.

The nodal account is typically provided by partner banks and audited for compliance with RBI settlement rules. Banks also provide payout APIs for transferring funds to merchant accounts.

---

*'Ministry of Home Affairs Press Notice', *Press Information Bureau*, 28 October 2024, https://www.pib.gov.in/PressReleaseIframePage.aspx?PRID=2069000. Accessed on 25 May 2026.

The scammers' first key insight was simple: any bank account can function as a nodal account if it is under their control—capable of receiving funds (via NEFT, RTGS, UPI, ATM deposits, cheques) and sending them onward.

Their second insight was that transaction tracking—without bank APIs—is still possible if they control these accounts via mobile phones running UPI apps.

Their third innovation was operational: deploying an army of operators using custom Android apps that instructed them to transfer funds at specific intervals in exchange for incentives. This effectively simulated settlement APIs.

Their fourth innovation was scaling: acquiring large numbers of such bank accounts by bribing officials or recruiting ordinary citizens as money mules.

Their final optimization was to get rid of the various payment options a payment gateway provides (e.g., credit card, debit card, net banking) and instead focus exclusively on UPI because of its widespread usage.

QR Codes and Dynamic Collection

UPI operates through aliases—recognizable payment addresses (e.g., user@bank) instead of complex account details (IFSC, account number). When converted into QR codes, these can be scanned by any payment app to initiate transactions.

Legal payment aggregators assign each merchant a unique payment address (e.g., merchant@fintech) and program their back-end rails so that when a user pays to this address, the callback from NPCI/bank returns the correct address and amount. They then register this payment on their ledgers and mark it as 'payable to the merchant'. Because of the escrow/nodal account design, the money from the user gets credited to the aggregator's escrow account, not directly to the merchant.

This enables aggregators to manage hundreds of thousands of

QR codes linked to a single nodal account. Each QR code identifies a specific merchant, even though all funds are pooled centrally.

The scammers' brilliance was in recognizing that users only needed to see a QR code. If the underlying bank account is known at a given point in time, transactions can be tracked and attributed. In their model, each loan app functions as a 'merchant' mapped to a pool of mule accounts, with associations maintained in a separate database—effectively replicating NPCI's mapping system.

Open-source intelligence and official reports suggest that retail savings accounts were initially used for collections, as loan repayments could be disguised as P2P transfers. Funds were then layered across multiple such accounts via UPI before being aggregated into current accounts held by shell companies.

These current accounts functioned as aggregators. From there, funds were moved out via NEFT and RTGS (rather than UPI), routed through additional layers of accounts, and eventually converted into cryptocurrencies or cash equivalents (e.g., gift cards, bill payments, e-commerce purchases).

Tasking and Moving Money at Fixed Cadence

Legal payment aggregators are required to settle money collected on behalf of merchants at a fixed cadence, typically referred to as T+1 or T+2 (a one- or two-day settlement cycle), with banks providing payout APIs to enable this. Scammers, however, operate very differently—they must move money constantly from one account to another, often within a 15-minute window, to evade law enforcement.

The basic premise behind this constant movement can be understood through the analogy of beaver tunnels. Imagine an underground tunnel system where each bank account represents a safe space. These spaces can be connected to others for short periods, but only by the beaver, as it knows how to navigate these two spaces. Now, when chased by a predator (in this case, law enforcement), the beaver's only reliable survival strategy is to keep moving from one safe space (bank account) to another.

While, in theory, movement is possible from one bank account to any other bank account, proximity and familiarity create natural clustering. The beaver prefers known paths because of ease of use, just as money tends to move along established networks. Because the predator can't be present in all spaces simultaneously and is always chasing the beaver, the beaver tunnel effectively becomes a four dimensional space.

Given that law enforcement is inherently slow, and the fact that only the beaver knows how to navigate the tunnel (as its architect), timed movements—such as shifting funds every 15 minutes—all but guarantee a very low chance of the predator catching the beaver.

Scammers implement this fundamental insight through 'movers', individuals tasked with transferring money between accounts under their control. Once funds reach an account, the mover is incentivized to pass them onward to another account—either pre-configured or assigned dynamically. This process is managed through 'tasking apps', which notify movers of transfers that must be completed within a fixed time window (typically 15 minutes) to earn incentives.

It is immaterial whether these accounts are opened using the movers' own credentials or sourced through corrupt bank officials. Since these tasking apps are not available on official app stores, they are side-loaded via APK files. In effect, movers resemble employees in a corporate network running proprietary software on controlled devices—except that the organization is a criminal enterprise.

Emergence of Money Laundering as a Service (MLaaS)

The systems built to extract income via fake loan apps and launder it are best understood as digital–financial ecosystems layered on top of existing infrastructure—KYC systems, bank accounts, payment aggregators and payout APIs. These systems are not single-purpose; they are reusable frameworks for moving money.

For instance, consider what the system is holistically:

- It allows money to enter through mechanisms such as under-invoicing of trade and parallel settlement systems.
- The money is then rotated across multiple layers of operators, each taking a cut.
- Finally, it is moved out of the country through various channels.

The IN–ROTATE–OUT model is highly adaptable and can support any business, as long as it:

- Is capital-intensive.
- Generates returns through digital channels (apps and web services).

Real-money gaming (RMG) and betting apps fit this model perfectly, as do cybercrime operations.

These businesses can also be understood as 'emotional dopamine triggers'. Take RMG for instance: it rewards users with frequent small wins (most participants recover some money), but losses, when they occur, are large and concentrated. This resembles selling naked options in financial markets—steady small gains punctuated by occasional catastrophic losses. The intermittent rewards keep users psychologically hooked, much like addiction. They know it is harmful, but really can't stop.

As documented extensively* by journalist Arti Singh, both legal and illegal gaming platforms use the same underlying techniques—dynamic QR codes, mule accounts, payout APIs—to

*Singh, Arti, 'The Shadow Economy of Online Gaming: How Money Moves beyond the Screen', *The Head and Tale*, 4 September 2025, https://theheadandtale.com/deepdives/the-shadow-economy-of-online-gaming-how-money-moves-beyond-the-screen. Accessed on 25 May 2026.

create a parallel financial ecosystem. Estimates suggest that as much as 30 per cent of UPI transaction volume may be linked to gaming and betting activities.

A payment executive quoted in her investigation captured the essence of the problem: 'UPI was meant to democratize payments. But it also democratized money laundering. Instead of one shady ₹20 lakh transfer, you now have two lakh ₹100 transfers. Good luck stopping that.'*

This refers to the 'split payments' feature. In legitimate contexts, such as e-commerce marketplaces, money collected from buyers is split—commissions are retained, and the rest is passed to sellers. MLaaS replicates this logic in the shadow economy. By breaking a large illicit transaction into thousands or millions of small transfers and routing them through vast networks of accounts and intermediaries, it becomes nearly impossible to distinguish laundering from legitimate activity. Cybercrime—where victims are conned out of their savings via emotional products like digital arrests—is one of the few visible indicators of MLaaS at scale, because its victims often complain to law enforcement agencies, triggering investigations, and hence some nodes in this 4D space end up getting busted and decommissioned. In contrast, gaming and betting ecosystems often operate as 'victimless' crimes, leaving much of this infrastructure invisible.

Weaponizing Law Enforcement Against Citizens

Scammers also deliberately plant misleading digital trails that implicate innocent people, who then end up facing harassment. Consider the case of Mrs H discussed in the first chapter. A scammer paid a taxi driver using a laundering-linked account, and later that day, the driver made a small payment to Mrs H.

As investigators followed the trail of split transactions, her account was flagged as a potential accomplice. Such aggressive

---

*Ibid.

enforcement has led to 'zero-trust-on-digital' environments in some areas. For example, trader associations in Nagpur reportedly stopped accepting UPI and card payments[82] because accounts were frequently frozen or lien-marked—often incorrectly—due to such indirect links.

As these cases increase, courts have begun urging law enforcement to exercise restraint in lien-marking. While this reduces collateral damage to citizens, it also allows scammers to retain more of their proceeds.

Thus, creating distrust in digital payment systems—while simultaneously exploiting them—is part of the criminal playbook.

In that respect, the behaviour mirrors corporate lobbying. Just as firms may shape regulations that hurt their competitors, scammers manipulate transaction systems to produce similar outcomes to route money using UPI—by bending regulatory systems to their will, not by influencing policymakers.

The 'genuine' transactions they generate are therefore no different from corporate lobbying in terms of outcome.

The primary factor that enabled scammers to build a multi-purpose money-laundering network is the lack of state capacity. The Indian state has several structural and embedded weaknesses, well documented elsewhere (e.g., it excels in mission mode and can coerce institutions and businesses to meet specific targets). To work around these weaknesses, it often avoids quantifying or measuring them rigorously. As a result, short-term, fixable problems accumulate alongside long-term structural issues that are deeply embedded and difficult to dislodge.

Fraud reporting is one such area. Journalist Aarti Singh's estimate places fraud at a very high upper bound (30 per cent), but it lacks methodological rigour and is closer to a gut estimate. At the other extreme is the regulator (RBI), which reports fraud at around 0.15 per cent.[83] How then can these two extremes be reconciled?

A few data points explain why the lower estimate is unreliable.

First, consider the methodology. The data comes from operators (banks, fintechs, wallet providers) as reported in the Central Payment Fraud Information Registry. However, when a ₹20 lakh fraudulent transfer is split into 2,00,000 transactions of ₹100 each and routed across multiple accounts, not all of these are reported as fraudulent. Yet each transaction contributes to total payment volume and value. Thus, the 0.15 per cent figure is not a reflection of reality but an artefact of the reporting process.

The I4C tracks[84] fraud indirectly by publishing the amount marked as a lien in accounts through which fraudulent money has passed. For 2023, the lien-marked amount was ₹921.59 crore, representing 12.32 per cent of total fraud, implying a total defrauded amount of ₹7,480 crore. RBI's payment system indicators for the same year report fraud at ₹3,835 crore[85]—an under-reporting of roughly 50 per cent. This suggests that fraud is at least 0.3 per cent of transaction value.

In 2024, I4C reported fraud of ₹22,845.73 crore,* while RBI's data showed only ₹5,297 crore†—a fourfold under-reporting. This implies a minimum fraud level of 0.6 per cent of transaction value. Given that at least one in two[86] victims does not report fraud, the effective lower bound rises to around 1.2 per cent of transaction value. Even this excludes other forms of laundering (e.g., real-money gaming, betting scams) and captures only cybercrime-related fraud such as pig butchering, task scams and investment fraud.

I4C reports only lien-marked amounts and does not track how much money is ultimately returned. Moreover, lien marking does not guarantee that funds have not already been withdrawn. This gap

---

*Government of India, Ministry of Home Affairs, *Lok Sabha Unstarred Question* No. 344, 22 July 2025, https://xn--i1b5bzbybhfo5c8b4bxh.xn--11b7cb3a6a.xn--h2brj9c/MHA1/Par2017/pdfs/par2025-pdfs/LS22072025/344.pdf. Accessed on 25 May 2026.

†Reserve Bank of India, *RBI Payment System Indicators*, December 2024, https://rbidocs.rbi.org.in/rdocs/PSI/PDFs/PSIDECEMBER20244E7A869B47184D939BD43C3D6E2CD930.PDF. The values from January 2024 to December 2024 in Table 5 were added together to arrive at the final figure. Accessed on 25 May 2026.

allows us to estimate how effective the laundering system is. If one in two victims does not report fraud, the lien-marked share rises to roughly 12 per cent in 2024 and 6 per cent in 2023. Assuming that 10 per cent of lien-marked funds are recovered, the loss to fraudsters is around 1.2 per cent in 2024 and 0.6 per cent in 2023.

If proceeds from cybercrime are treated as 'capital', analogous to the 'assets' of financial institutions (e.g., banks, NBFCs), then losses to fraudsters are equivalent to non-performing assets (NPAs). At 0.6 per cent in 2023, this is comparable to top-tier banks (average NPAs ~0.4 per cent). At 1.2 per cent in 2024, it is in line with leading NBFCs.

That a criminal enterprise can generate such returns—while operating under pressure from regulators, law enforcement and financial institutions—offers a sobering insight: money-laundering networks are not only capital-efficient but also organizationally more efficient than many parts of the formal financial system.

This lack of control illustrates how systemic failures accumulate over time. When left unaddressed, they give rise to parallel systems, powered and amplified by digital technologies. Monitoring digital supply chains should therefore be central to financial regulation and law enforcement. Yet these institutions often lack the holistic perspective required to understand how interconnected these systems are.

In contrast, market-driven fraud networks—often aided by cross-border linkages—are able to exploit these interconnections with ease.

Mastery of digital supply chains in even one domain thus provides a decisive advantage across multiple domains.

6

A Freedom-Fuelled Future

Let us explore a question that demands more than a cursory answer.

What does it mean to be a technology superpower when you do not own the tools, cannot shape the rules, and are excluded from writing the code that governs the world?

Imagine a grand temple, layer upon layer, rising from the earth. At its base lie chips—the silicon hearts that pulse to the rhythm of our modern world. Above them come compilers and software toolchains, shaping raw logic into structured instructions. Frameworks stand even higher, the scaffolding that maps the terrain of artificial intelligence. At the top sit large language models, vision networks and conversational systems. And finally, there are apps—the gleaming temple doors through which users pass every day. India today polishes those doors far more than it builds the other layers. The sanctum sanctorum and its foundations are built elsewhere—with our labour, but without our ownership or control.

Walk into an office park in Bengaluru on any weekday, and you might see 20-something engineers poring over chip designs that will eventually power devices in a German car, an American fighter jet or a Chinese drone. Their screens flicker with microarchitectures and simulation outputs, yet none of these chips will ever bear an Indian trademark. We built them. But they belong to someone else. That is the quiet tragedy of our digital economy.

We design chips in Bengaluru and Hyderabad. Our engineers write code that flows into AI hardware across the world. Yet they do so for foreign firms. Elsewhere, nations build sovereignty into their architecture. China funds chip fabrication. The United States backs DARPA and the CHIPS Act. Taiwan nurtures TSMC. India hosts hackathons. We celebrate startup awards. We tweak quotas. But without infrastructure, there is no sovereignty—only dependency cloaked in loud narratives, often propagated breathlessly on WhatsApp.

There is excellent engineering, yes—but it is individualistic, not societal.

That asymmetry matters.

It raises uncomfortable questions.

Why are we so willing to be value creators for others, but not for ourselves? Why are we satisfied with individual wealth, even when it condemns us to collective poverty? Why do we accept a role where economies of scale flow outward, patents are filed abroad, and our platforms pay licence fees to global giants?

We have an education system that celebrates enrolment but often lacks depth. China produces multiple times our STEM patents. Our top talent migrates because ambition demands research, facilities and purpose. Abroad, engineers are funded, granted resources and given a voice. In India, they build prototypes—while their ideas take flight elsewhere. Our R&D spending, especially at the academic and corporate level, remains pitifully low. Indeed, many of our most successful innovations—from space missions to digital payments—have emerged from public investment and public institutions.

We speak of digital sovereignty, yet we import the very bricks on which it depends. We aspire to sovereignty while buying chips from foreign fabs, licensing compilers and storing public data in external clouds. Sovereignty must begin from the ground. It must reside in hardware, in compilers, in open-source ecosystems. And it must scale to global markets because the world rewards ubiquity.

History offers a clear precedent. The subcontinent once accounted for a large share of global trade in cotton textiles, pepper and silk. Its goods traversed global shipping lanes. Then came industrial-era isolation, flawed policy and inward-looking priorities. We protected markets rather than built industries. We mistook quotas for capability, planning for execution. But scale creates margins and innovation. It exposes inefficiency. It enforces standards. It builds ecosystems. If we do not build for scale, we do not build at all.

That is why strategic digital trade matters. An Indian open-source LLM cannot remain an internal prototype or a symbolic exercise. It must perform at scale across diverse global environments—Arabic, Swahili, Spanish and beyond. It must be scrutinized, co-created and interoperable. It must earn trust through use—not through restriction or token gestures of 'swadeshi' labelling.

Trade is not an exhibition. It is a web of dependencies. When nations adopt our chips or models, they embed us into their economies, their policies, their politics. They bind our code with their citizens' rights. That creates strategic presence and influence—not superficial clout, but structural integration into global systems.

India has talent, a demographic dividend in theory, and perhaps even a democratic DNA. But talent without tools is latent energy leaking away. We must build foundations—fabrication labs, open compilers, long-term research funding that spans decades, not quarters. We must treat digital policy the way Venice built canals, Japan built cars, and Germany built engineering institutions: patiently, systemically, over generations. If we cannot think in decades—and if we substitute ambition with slogans of having already 'arrived' as a *Vishwaguru*—we will not build lasting capabilities.

A 50-year vision is not optional. In capitalism, entrepreneurs often cannot think beyond exits. States must. If they cannot, one must ask what purpose they serve. This requires public investment that tolerates risk. It requires a trade policy that rewards scale rather than protects inefficiency. It requires moving from being vendors to becoming builders.

The alternative is bleak. We become the backend to other nations' digital revolutions—our engineers disconnected from the platforms they imagine. We implement algorithms but do not write them. And when supply chains shift, or geopolitical tensions rise, we find ourselves locked out—unseen, yet strategically stranded.

Yet we can choose otherwise. We must. It is time for both competence and conviction. It is time to build the digital temple properly—from chips beneath our soil, to toolchains in our hands, to models shaped by our insights, to systems trusted by other nations. We must be the builders, not merely the users.

India must treat an AI-driven future not as a defensive posture, but as proactive trade policy—not to barricade AI, but to import ideas and frameworks, absorb them, master them, contribute to them, and export them globally at near-zero marginal cost.

What a Freedom-Fuelled Future Means

Imagine the year 1954, when both India and China agreed to the Five Principles of Peaceful Co-existence.* These principles were later endorsed at the Bandung Conference (1955) and became a cornerstone of the Non-Aligned Movement (NAM). They were seen as a model for international relations based on peace, equality and mutual respect—especially relevant for countries emerging from colonial rule—and served as a framework to navigate the geopolitical environment of that era.

2026, however, is not 1954. There is uncertainty all around. In terms of industrial capacity and innovation, China is no longer seen as a near peer of the United States; instead, in many areas, the United States is now viewed as a near peer of China. As a yet to emerge middle power that has largely favoured the US, yet is geographically bound to China, India needs a new set of principles

*Government of India, Ministry of External Affairs, *Panchsheel*, External Publicity Division, https://www.mea.gov.in/uploads/publicationdocs/191_panchsheel.pdf. Accessed on 25 May 2026.

to build leadership in disruptive digital technologies such as Artificial Intelligence.

We call these the Five Principles of Digital Panchsheel.

Build for the World: First, India must create export-linked incentives for AI companies. Think of it as an SEZ model for the new era—where companies that release open-source models, contribute to the global AI commons, and enable sovereign stack options for friendly nations receive priority access to grants, tax breaks and compute infrastructure.

Embrace Openness: Second, all national AI initiatives must be released in the open. No more secretive, vendor-built, closed-loop LLMs that claim to represent 1.4 billion Indians. Open source is not just good ethics—it is good economics. It enforces quality through global peer review and avoids vendor lock-in. The moment one shuts the door to outside critique, mediocrity follows.

Govern with Wisdom: Third, regulators must stop confusing caution with paralysis. Yes, AI misuse must be regulated—against bias, discrimination and surveillance—but laws must evolve with technology. The only thing worse than unregulated AI is frozen innovation. Paranoia about what is not well understood cannot become policy.

Retain Agency and Independence: Fourth, companies must resist being driven purely by government contracts. India's IT industry succeeded because it did not become a public-sector vendor ecosystem. Infosys, TCS and Wipro became global by solving real business problems, not by customizing portals for ministries. AI companies must avoid the seductive GovTech trap and resist being pulled into geopolitical 'choose-your-side' dynamics.

Invest in Commons and Diverse Algorithms: Finally, India must create a public digital talent commons. Just as the Green Revolution was built on agricultural universities, and the telecom revolution on

IIT-trained engineers, India must now fund AI fellowships, establish AI labs in major universities, build vernacular NLP testbeds across Indian languages, and explore bold alternatives—including beyond transformer architectures. This is the infrastructure of thought.

- In a world increasingly shaped by John Mearsheimer's realist framework from *The Tragedy of Great Power Politics*, these principles offer a way to navigate the emerging AI Cold War. Mearsheimer's offensive realism argues that: Great powers are inherently driven to seek dominance.
- The international system is anarchic, with survival as the primary goal.
- Even reluctant states must arm and align, because others will.
- Middle and smaller powers survive through balancing, banding together or strategic neutrality.

In today's AI landscape, the US and China are the two superpowers, each advancing its own techno-political ecosystem. The world is fragmenting into digital spheres of influence, where domestic companies are increasingly aligned with geopolitical agendas. Middle powers such as India and Brazil face Mearsheimer's dilemma: independence is risky, yet alignment erodes sovereignty.

Mearsheimer would argue that freedom and openness are luxuries of the powerful. A middle power must align, balance or risk being dominated. Therefore, a freedom-fuelled future can only be a realist strategy if 'freedom' translates into autonomy and power—not merely ideals. By building full-spectrum sovereign AI supply chains, national compute infrastructure and public digital capacity, a nation creates the conditions necessary to avoid dependency and withstand the weaponization of interdependence.

'Fuelled' also implies being resourced, trained and prepared—not naïve. AI fellowships, open models and domestic R&D build national capability. State-supported public goods—compute, data and language models—make coercion more difficult for external powers. While offensive realism focuses on short-term survival, middle powers must play a longer game—building compounding advantages over time. 'Future' thus reflects not optimism about global goodwill, but a clear-eyed recognition of shifting power dynamics and the need for sustained strategic adaptation.

Building for the World: Avoiding State+ and Status++ Incremental Innovation

Even now, as the world reshapes itself at the pace of algorithms and the scale of ecosystems, India often falls back on the comfort of incrementalism. It is a deep-rooted reflex—not born of incompetence, but of an entrenched habit of taking what exists, adding a local flavour and calling it sovereignty. Our approach to artificial intelligence today is a prime example of what may be called a 'current status-plus-plus' model, combined with half-hearted state support—a 'state-plus' approach. It is cautious, conservative and deeply out of sync with how true innovation works.

We look at today's digital components—built largely elsewhere—and ask how to localize them, without understanding the depth of the supply chains that power them. We imagine closed-loop large language models trained on government-provided data, built by selected vendors and housed in domestic cloud silos. We ask whether we can build an Indian version of ChatGPT, or a Hindi-speaking assistant for government portals. These are not foolish questions—but they are the wrong place to stop. At best, they move us from dependence to decoration. At worst, they entrench irrelevance under the illusion of control.

The problem lies in mistaking sequential progress—measured through activities and timesheets—for exponential transformation. True systemic innovation is not serial. It does not move in neat steps. It leaps. It reinvents the problem even as it solves it. Consider social media. It did not exist, even as a concept, two decades ago.

No policy planner in 2000 could have anticipated a world where a teenager in Kerala could become an influencer, or where elections could turn on the swing of a digital narrative. Social media was not an upgrade from television; it was a different world altogether.

This is what the status-plus mindset misses. It assumes the future is a modified version of the present—and seeks state support to reinforce it—thereby locking mediocrity into policy, since the state's natural tendency is to preserve the status quo. But the future, especially in technology, is often unrecognizable.

AI will not merely automate existing processes; it will refactor entire sectors. In healthcare, it will not just digitize records—it will redefine diagnosis. In education, it will not simply deliver lessons—it will reshape how knowledge is structured and shared. In governance, it will not only accelerate services—it will shift the axis of accountability. These are not upgrades; they are rewrites—compressing the evolution of human systems themselves.

And yet, we continue to act as if AI were a compliance checklist—another item in a bureaucrat's promotion pathway. We seek national platforms without articulating a national purpose. We fund sandbox pilots without clarity on global positioning. We draft regulations that treat AI as a risky tool rather than a strategic asset. We protect domestic datasets without asking whether they are usable, intelligent or even of sufficient quality.

The result is policy theatre. The noise is loud, the activity visible—but outcomes remain unchanged. The script is familiar, the actors competent, but the stage has already shifted.

This is not how China built Huawei, or how the United States backed OpenAI, or how Europe used GDPR not just as a regulation but as a tool of power positioning. These are not status-plus moves; they are paradigm shifts. They do not accept the architecture of the world—they attempt to redesign it and export that architecture globally.

It is no longer enough to ask: 'What is the Indian version of

X?' The real question is: 'What can India offer that the world has not yet seen?' In the AI century, relevance does not come from imitation; it comes from introducing the unexpected. That requires vision, not just velocity. Without it, we risk becoming managers of someone else's future.

Building for the world—at least conceptually—offers a way out of this mediocrity trap. India's own IT services story from the 1980s shows how. The sector did not grow by relying on domestic demand; it went global, competed in hard currencies, delivered under pressure and learnt under constant scrutiny. These firms did not wait for regulatory protection or guaranteed markets—they pursued opportunity across borders.

Even today, the contribution of foreign clients to leading IT services firms remains extraordinarily high (e.g., Infosys ~97 per cent,[87] TCS ~94.4 per cent,[88] Wipro ~97 per cent,[89]). In theory, this should have driven investment in breakthrough technologies. Instead, the services model produced a poorly understood side effect—it bound firms tightly to their clients. The steady margins reduced incentives for risky bets, while public listing imposed quarterly performance pressures. Incrementalism became the norm, even as these firms operated globally—because the problems they chose to solve were themselves incremental.

Consider Infosys's early donation* to OpenAI in 2015, when LLMs were barely conceptual. The initiative was driven by then CEO Vishal Sikka, who also served as an adviser to OpenAI. When he exited in 2017 following disagreements with the board, that vision dissipated. In hindsight, it stands as a lesson in how even early insight can fail without institutional continuity—the ability to recognize a shift is meaningless without the ability to commit to it.

---

*Kannan, Uma, 'Infosys among Early Investors in ChatGPT Creator', *The New Indian Express*, 13 January 2023, https://www.newindianexpress.com/business/2023/Jan/13/infosys-among-early-investors-in-chatgpt-creator-2537418.html. Accessed on 25 May 2026.

Another constraint emerges in domestic government projects. These firms increasingly secure large public contracts (e.g., TCS[90]), often struggling* to deliver them while publicly acknowledging losses.† Why participate at all?

The answer lies beyond margins. What firms lose financially, they gain in political and policy influence. Leadership becomes embedded in advisory committees, shaping regulatory frameworks. This influence becomes a short-term asset—but a long-term liability. It reduces risk domestically, while embedding bureaucratic risk-aversion within corporate culture. The result: 'status-plus-plus' evolves into 'state-plus' across the ecosystem, as talent circulates between firms.

Global Competency Centers (GCCs) represent a natural evolution of the services model, with key differences. They provide direct exposure to parent organizations rather than mediated relationships, along with deeper technical engagement and pathways for global mobility. Yet, they too remain extensions—not origins—of innovation.

Truly global software products from India remain rare (e.g., Postman, Freshdesk, Zoho, BrowserStack), but they demonstrate a critical point: building products for global markets—not projects—significantly increases the probability of innovation.

AI, however, introduces new dynamics that challenge even the SaaS model. Training models is expensive and difficult, but using them is becoming easier—and will likely converge towards the cost of electricity.

---

*Srikanth, Chandra R., and Swathi Moorthy, 'Why Indian IT Struggles with India Government Projects', *Moneycontrol*, 21 June 2021, https://www.moneycontrol.com/news/business/why-indian-it-struggles-with-india-government-projects-7066591.html. Accessed on 25 May 2026.

†'Infosys Always Lost Money on Government Projects: Narayana Murthy', *Hindustan Times*, 21 July 2015, https://www.hindustantimes.com/business/infosys-always-lost-money-on-government-projects-narayana-murthy/story-Gg2yMUdck99LM9N9Hmk2TI.html. Accessed on 25 May 2026.

This shifts the strategic calculus. Giving capabilities away—especially to friendly nations—may become necessary to drive usage, research and ecosystem development. It is a bold and risky bet to assume that investments made by private players cannot be recouped through individual companies alone but only across the entire ecosystem and the applications that spawn around it. Yet, given the dynamics at play, there may be no choice but to make that bet and go for broke.

Embracing Openness: Not Masking Mediocrity through Patriotism

A famous Tamil proverb sums up India's peculiar situation: 'There is a stone, there is a dog, but there is no one to pick up the stone and throw it.' The absence of even one ingredient prevents outcomes. Only by acknowledging what is missing can one begin to address it. However, historically, our cultural instinct has been to pretend otherwise.

Consider the oft-repeated claim that India has abundant talent—without defining what 'talent' means in a rapidly evolving technological landscape. The Economic Survey 2023–24* notes that only one in two graduates is employable, and it is widely acknowledged that the education system is broken.† These deficiencies stem from a lack of industrialization, coupled with insufficient capital to invest in breakthrough and high-risk technologies.

The reforms of 1991 were incomplete and later stalled by the infamous Bombay Club, which argued that 'infant industries need protection from predators'. This delayed deeper reforms until 2004.

---

*'Only 51% Indian Graduates Job-Ready in Employability Crisis: Economic Survey', *India Today*, 22 July 2024, https://www.indiatoday.in/education-today/latest-studies/story/only-51-indian-graduates-job-ready-in-employability-crisis-economic-survey-2570363-2024-07-22. Accessed on 25 May 2026.

†Mandavia, Megha, 'India's Broken Education System Threatens Its Superpower Dreams', *The Wall Street Journal*, 21 April 2024, https://www.wsj.com/world/asia/indias-broken-education-system-threatens-its-superpower-dreams-19b77da6. Accessed on 25 May 2026.

The real surprise is that even these limited reforms enabled Indian firms to compete globally. Yet the underlying fear persists. In the digital era, it reappears as tariff barriers, import substitution and an enduring obsession with self-reliance.

The result is not merely the absence of action—it is the absence of fundamentals. A large segment of the young population remains unemployable. Many graduates earn stagnant starting salaries for nearly a decade—barely enough to make ends meet*—while being exhorted to work 70 hours a week[91] for nation-building.

Against this backdrop, India now seeks to build world-class AI, semiconductor ecosystems and digital infrastructure—using a funding base designed for incremental improvement, not moonshots. India did not build institutions like DARPA. It did not foster equivalents of TSMC, Bell Labs or the MIT Media Lab. Public research institutions were either bureaucratized or starved of resources. The private sector, long accustomed to captive domestic markets, had little incentive to invest in deep technology.

Venture capital arrived late—and often with time horizons too short for foundational innovation. In the absence of investment capacity, India has retreated into narrative. Markets are protected, products are branded as 'indigenized', and sovereignty is asserted rhetorically. But true technological sovereignty requires trusted competence and a culture of constructive scepticism. India has too little of both. Instead, it often believes its own narrative—mistaking insulation and irrelevance for independence.

Consider one of India's genuine strengths in the AI supply chain: a large and active open-source community. Spread across cities and domains, it contributes to major global projects such as PyTorch,

---

*'From 3.2L in 2011 to Just Rs 3.75L in 2024: Ex-Infosys CFO Mohandas Pai's Shocking Comparison between CEOs vs Freshers' Salaries', *The Economic Times*, 23 December 2024, https://economictimes.indiatimes.com/jobs/fresher/from-3-2l-in-2011-to-just-rs-3-75l-in-2024-ex-infosys-cfo-mohandas-pais-shocking-comparison-between-ceos-vs-freshers-salaries/articleshow/116592116.cms. Accessed on 25 May 2026.

ERPNext, Hoppscotch and Hasura GraphQL. These contributions are rigorously scrutinized, critiqued and commented on before adoption. They meet global standards without any 'protection' or allowance.

This is the same approach followed by Chinese researchers behind DeepSeek. The model is open-weight—not restricted behind APIs. Zhipu AI went further by open-sourcing its training process. Platforms such as LM Arena[92] consistently show open-source models ranking among the top 10 across tasks.

Yet the instinct to avoid scrutiny runs deep in India—driven by a fear of failure and of not being good enough from the outset. For example, Sarvam.AI, despite receiving taxpayer funding and privileged GPU access, initially avoided committing to open weights, doing so only after public criticism.*

To counter this instinct, a radical shift is required: India must commit to building fully open-source models with complete traceability, transparency and reproducibility from scratch—similar to EleutherAI's[93] *Pythia*. Only such openness offers a realistic path to catching up in AI, given the late start and structural disadvantages.

Govern with Wisdom

Two dominant instincts shape the Indian state's approach to technology. The older instinct is to define what is not allowed through licensing and permissions. The newer, post-liberalization instinct is to prescribe what must be done. This 'mandatory–not allowed' duality has become the default policy framework. In both cases, decisions are often made without a deep understanding of evolving technologies, making policy vulnerable to capture by special interests or charismatic actors.

*'Inside India's Scramble for AI Independence', *Maverick Studios*, 4 July 2025, https://maverickstudios.net/2025/07/04/inside-indias-scramble-for-ai-independence. Accessed on 25 May 2026.

Nowhere is this more evident than in AI regulation, particularly in AI ethics. Consider the nature of language itself. At any moment, an LLM generates sequences of words from a vast space of possibilities. Some outputs are undesirable—because they offend sensibilities or raise concerns such as national security.

The only way to constrain such outputs is through training on examples and counterexamples—effectively shaping a probability distribution. However, it is impossible for society to agree on all undesirable outputs, let alone provide exhaustive training data.

This leads to a fundamental reality: no matter how extensively an LLM is trained, there remains a non-zero probability of generating undesirable outputs—and this probability cannot be precisely estimated.

Instead of acknowledging this mathematical constraint, AI ethics and frameworks often devolve into risky behaviours, assigning human fallibilities* to what is essentially a probability distribution function. This anthropomorphism results in misguided regulatory approaches.

Consider NASSCOM's Responsible AI[94] playbook and its identified risks:

1. Unintended or malicious use by third parties
2. Generation of biased, dangerous or illegal outputs
3. Uncontrolled deployment

An open-source strategy—designed to accelerate research—necessarily enables broad, even uncontrolled, deployment. That is not a flaw; it is the intent. Similarly, preventing all harmful outputs is infeasible by design, as no exhaustive set of examples can exist.

In effect, policymakers attempt to impose human characteristics on a probabilistic system, treating a token-generation engine as if it

*For a more rigorous mathematical treatment of this aspect, see 'A Non-Anthropomorphized View of LLMs', https://addxorrol.blogspot.com/2025/07/a-non-anthropomorphized-view-of-llms.html. Accessed on 25 May 2026.

possesses agency. This is not merely over-regulation—it reflects a lack of state capacity to understand an evolving field, coupled with the illusion of control.

The consequence is predictable: in trying to regulate what it does not understand, the state risks stifling the very creativity required to build it.

Retain Agency and Independence

Consider a few recent incidents that illustrate how the distinction between a nation-state and companies operating within its boundaries is increasingly blurred:

1. As a precondition for its US$1.5 billion[95] investment, Microsoft required G42 to remove Chinese chips and equipment from its data centres and divest its holdings in Chinese companies.
2. Microsoft cut off the email access[96] of Karim Khan, the ICC prosecutor, following sanctions imposed by Donald Trump.
3. Ukraine lost access to satellite imagery* from Maxar, reportedly to pressure it into negotiations.
4. OpenAI's 'OpenAI for Countries', sold as 'Democratic AI', explicitly states† that the goal is to 'maintain the world's most advanced AI infrastructure in America', effectively aligning AI safety with US strategic interests.‡

---

*Haworth, Jon, Oleksiy Pshemyskiy, Patrick Reevell, David Brennan, Nataliia Popova, and Matt Seyler, 'Ukrainian Forces Lose Access to Maxar Satellite Imagery via the US', *ABC News*, 8 March 2025, https://abcnews.go.com/International/russia-conducts-massive-aerial-attack-ukraine-overnight/story?id=119543077. Accessed on 25 May 2026.

†OpenAI, *OpenAI for Countries: Our Approach to Security*, https://cdn.openai.com/global-affairs/ff86ec36-1741-40a0-ab2a-d46f8eec62a8/openai-for-countries-our-approach-to-security.pdf. Accessed on 25 May 2026.

‡Note that this is one reason why we are exceptionally sceptical about prevailing narratives in India on AI Safety. By taking an unsolvable—but only mitigable—problem of not being able to estimate or bound the probability distribution function of unwanted or undesirable

5. Palantir CEO Alex Karp has openly stated[97] that his company is dedicated 'to the service of the West, and the United States of America, and we're super-proud of the role we play, especially in places we can't talk about.'

This fusion of private commercial interests (to maintain market share and domination on a technology) with state priorities is not new. The closest historical parallel in the Global South is the British East India Company. China follows a similar model today, where private firms such as Huawei and Ant, alongside state-owned enterprises, are treated as extensions of the state—mobilized to serve national priorities and goals (whatever they may be).

If an Indian company were to build a foundation model and a broader AI ecosystem, it would need to differentiate itself by not being tightly bound to the priorities of the Indian state. Otherwise, it risks mirroring the US–China model without possessing comparable capabilities—ensuring that global users prefer the more advanced offerings from those ecosystems.

This is already a potential challenge for companies such as Sarvam.AI. Reports suggest that in exchange for priority access to GPU infrastructure, the government may take[98] an equity stake. While this may accelerate initial progress, it also risks aligning the company too closely with state priorities. Even if technically successful, such alignment could create scepticism in global markets, limiting adoption.

The other problem of becoming too tied with the government ecosystem is that it exposes the companies to familiar systemic weaknesses: slow execution, bureaucratic processes, shifting policy priorities and narrow use cases driven primarily by government

---

word sequences, and framing it as AI Safety, which is also the explicit framing used by OpenAI to align this with US geopolitical interests in maintaining AI dominance, and considering existing examples of coercion through the use of kill switches to force countries to toe its line, we believe that the US does not believe in sovereignty (representing agency) at all. India would therefore be extraordinarily naïve to fall into this trap.

demand. Some of these issues are already visible. For example, apart from Sarvam.AI, most startups have not yet received GPU[99] access. Even before such allocation is resolved, the policy focus appears to be shifting towards supporting Small Language Models (SLMs), potentially diluting strategic clarity.

Invest in Commons and Diverse Algorithms

At an abstract level, culture is analogous to DNA—both encode accumulated history. As LLMs become mainstream and the primary interface for accessing knowledge, cultures not represented in training data risk gradual erasure. If that culture is embedded in local languages and mediated through a dominant language such as English, multiple layers of lossy compression compound the loss.

This makes the creation of local-language corpora essential. Initiatives such as Viswam.AI (originating from Swecha[100]) have taken early steps—digitizing thousands of stories, including culturally significant works like *Chandamama Kathalu*. The effort has expanded to include folk tales, idioms and lived experiences contributed by volunteers across Telangana and Andhra Pradesh, creating a rich and diverse textual base.

In theory, this dataset may turn out to be the only defensible moat that Indian startups can build. But this raises a critical question: how does one establish that a given dataset belongs to the Telugu community, and ensure it cannot be claimed by a foreign government or company? Conceptually, this problem is no different from geo-tagging basmati rice through a Geographical Indicator. If Telugu culture is considered indigenous to India—and specifically to Telangana and Andhra Pradesh—then state-funded corpora and LLMs trained on them should have the right to assert a cultural indicator. Creating the institutional framework to recognize, claim and enforce such cultural indicators is a common need across the Global South.

Viswam.AI also demonstrates another approach to building the societal infrastructure required to create a local language LLM. It launched a Summer of AI project to train 1 lakh engineers on AI Skills,[101] though the progress has been hard-won.[102] This signals an alternative model of AI development—one that throws passion at the problem, rather than only throwing millions at it.

It is also researching alternatives to the Transformer architecture, in part to avoid the capital-intensive GPU arms race that transformer models demand. While one might be tempted to dismiss this as a wild-goose chase, Meta's researchers, when asked, agreed that this is the need of the hour—acknowledging that even they are internally exploring alternatives, for the following reasons:

- Transformers are primarily based on language.
- Human intelligence, however, is highly multi-modal—suggesting that a truly capable AI system needs to be multi-modal as well.
- Transformers, however, output discrete tokens that are unimodal in distribution.

Backed by a state government with close to zero funding from the central government, Viswam.AI also points out that diverse, grassroots strategies may outperform the conventional approach of only throwing more GPU and money—or ceding control to centralized government allocation of scarce resources, which risks unconsciously replicating the license raj (e.g., public dataset, community participation, alternative architectures, shared compute on phones).

Economic Freedom as the First Principle of Digital Sovereignty

For any nation aspiring to digital sovereignty, the first step is not code, policy or infrastructure. It is freedom—beginning with economic freedom. Sovereignty, at its core, is not a technical condition; it is a condition of confidence. Only when an economy can stand on its own feet can it build technologies on its own terms. Without that foundation, digital sovereignty becomes a rhetorical ornament—fragile in crisis and dependent by design. This is why economic freedom must precede digital freedom.

Economic freedom gives society the room to make mistakes, invest patiently, and absorb shocks. It allows entrepreneurs to experiment without fear of bureaucratic reprisal. It gives citizens the ability to demand accountability. And, crucially, it gives the state the capacity to invest long-term without constantly seeking foreign capital or validation.

In a world where digital power is concentrated in a few global corporations, nations that lack economic resilience become digital clients, not digital architects. They adopt platforms they cannot regulate, outsource systems they cannot secure, and depend on data centres they do not own. Their public infrastructure ends up subsidizing foreign balance sheets. Their innovation energy is channelled into compliance, not creativity.

Economic freedom, therefore, is not merely about deregulation or lower taxes. It is about enabling distributed energy—where thousands of local experiments can coexist. When local economies thrive, digital ecosystems follow. A society with strong economic roots can afford to build sovereign digital moats—systems that protect its data, talent and innovation cycles from being drained away.

The Bureaucratic Paradox

India's bureaucracy was designed for command and control. Its foundational logic is one of surveillance and supervision—a structure inherited from a colonial state, where administration meant extraction and compliance. The paradox is that while individual officers often display extraordinary initiative, the system itself discourages systemic innovation. There is little incentive for a bureaucrat to go beyond the baseline in delivering citizen outcomes. The 'stick' in the carrot-and-stick framework dominates, suppressing initiative before it can emerge.

A young district collector begins as a de facto CEO of a district—responsible for health, education, agriculture, law and order, and disaster management. In the private sector, such a role would come with autonomy, flexible teams and measurable outcomes. In government, it comes with paperwork, overlapping jurisdictions and constant audit anxiety. As officers rise through the ranks, they move further from problem-solving and closer to file-clearing. The career arc shifts from entrepreneurship to caution.

Contrast this with China's bureaucracy. For all its flaws, it has treated local administrators as economic actors. A city party secretary's advancement depends on measurable outcomes—GDP growth, employment generation, investment attraction and infrastructure development. Incentives are aligned with economic performance, which in turn shapes social and political standing. Chinese bureaucrats function, in effect, as venture managers—mobilizing local capital, attracting industry and delivering visible growth.

This is not to romanticize China's model—it carries risks of authoritarian overreach and systemic opacity. But its core lesson over four decades is clear: economic autonomy at the local level translates into national strength and geopolitical power. When local administrators have both resources and responsibility, they become entrepreneurs of governance. When they are reduced to custodians of files, they become bottlenecks.

Building Sovereign Ecosystems, Not Dependent Supply Chains

India's digital transformation has outpaced its economic decentralization and, in many ways, worsened it. Many of the digital systems have relied on strong central backing, often enforced through fiscal and administrative pressure. Cities depend on grants from Delhi. Startups depend on capital from California. Manufacturing depends on components from China. Data infrastructure depends on foreign cloud providers.

This weakens sovereignty. When the economic base is externally dependent, digital independence becomes performative. We may write our own code, but we rent the servers. We may regulate foreign apps, but rely on foreign capital. We may promote startups, but benchmark them against external valuations.

True sovereignty emerges from self-sustaining ecosystems. Economically strong localities generate taxes, talent and trust—the three foundations of autonomy. They can fund research, build infrastructure and negotiate externally from a position of strength. Without this base, even well-intentioned digital policy remains defensive—a shield rather than a strategy.

Cities as Engines of Freedom

If India seeks digital sovereignty, it must begin by empowering its cities. Cities are where physical and digital life intersect—where data is generated, businesses grow and services are delivered. Yet Indian cities remain financially constrained. Most municipal budgets depend on state transfers. Property taxes are outdated, user charges are politically restricted, and borrowing frameworks are weak.

The result is predictable: cities lack the authority to shape their own economic and digital futures. They cannot invest in innovation or infrastructure without higher-level approvals. Leadership is administrative rather than political; citizens are treated as clients rather than stakeholders.

Imagine cities as sovereign economic units—able to raise bonds, attract investment, manage data trusts and collaborate with startups. Imagine cities experimenting with infrastructure the way states experiment with policy. That is how sovereign ecosystems emerge—from empowered localities that control both economic destiny and digital identity.

But empowerment without integrity creates new dependencies.

Cities cannot become instruments of electoral extraction—drained during campaigns and neglected afterwards. When governance becomes patronage, cities cease to function as public institutions and instead become marketplaces of favour and corruption. No digital strategy can succeed in such an environment.

If India seeks an inclusive future, it must protect cities from capture—financial, political and moral. Transparency must extend across budgets, contracts and data flows. A city that cannot resist corruption cannot enable innovation.

India must fund its cities not as dependents but as partners—through predictable, performance-linked transfers, coupled with accountability that rewards innovation and penalizes stagnation. But funding alone is insufficient; cities need autonomy.

Economic freedom at the city level also demands cultural change. Bureaucrats must evolve from controllers to enablers, from auditors to ecosystem builders. Public institutions must attract talent from technology, design and behavioural sciences—not just administrative streams. Governance must become entrepreneurial: agile, risk-tolerant and citizen-centric.

When cities operate with such freedom, digital ecosystems emerge organically. Universities link with industry, startups address civic challenges, citizens co-create services, and data becomes an asset rather than a liability. The city becomes both laboratory and marketplace—a self-sustaining digital polity.

Economic freedom is not a libertarian indulgence; it is strategic insulation. Economically weak nations must compromise digitally—they buy rather than build, conform rather than negotiate. Economically confident nations can choose independence.

The ability to fund indigenous innovation, retain intellectual property, and scale domestic manufacturing flows directly from economic autonomy. Without it, digital sovereignty is reactive—a scramble to keep pace. With it, sovereignty becomes proactive—a framework to shape the rules.

Economic freedom builds confidence; confidence builds capacity; capacity builds sovereignty.

To achieve this, India's governance philosophy must shift from control to trust—trust in markets, cities, local governments and citizens. This does not imply the absence of regulation, but the presence of respect. The state must evolve from a parent into a partner.

This transformation demands a new kind of leadership—administrators who think like investors, investors who think like citizens, and citizens who see governance as a shared enterprise. It requires policy courage: to devolve fiscal power, simplify compliance and protect risk-taking. And above all, it requires imagination—to recognize that digital sovereignty is not an IT project, but a long-term social contract between citizens and the state.

Coda: Digital Fairness and the City as Custodian of Sovereignty

Digital fairness is not an abstract moral ideal; it is the visible outcome of how power, opportunity and accountability are distributed across a society. A nation that concentrates decision-making in a few offices cannot expect its digital systems to feel inclusive. Fairness in the digital century begins where fairness in governance begins—at the level closest to citizens.

Cities, therefore, will become the true custodians of digital sovereignty. It is within them that identity, data, services and innovation converge. If a city can protect the dignity of its citizens while enabling innovation, if it can ensure that digital infrastructure serves the public rather than exploits it, then the republic's sovereignty becomes real, tangible and participatory.

Economic freedom gives cities the autonomy to pursue this fairness. A financially empowered city can invest in public digital infrastructure that reflects local needs rather than imported templates. It can build open-data ecosystems where citizens, researchers and startups collaborate without fear of bureaucratic reprisal. It can pilot ethical frameworks for AI and automation aligned with community values. None of this is possible if cities remain dependent, cash-strapped and voiceless.

Fairness also requires transparency—a culture of openness where digital systems are audited, policies are debated, and outcomes are visible. When citizens understand how decisions are made, they trust them. When they can see where their taxes and data go, they participate more willingly. Digital fairness, in this sense, is not a feature of software; it is a function of citizenship.

The task before India's policymakers is to trust its cities and citizens with greater freedom, not less—to allow them to shape their own economic ecosystems, experiment with digital governance, and adapt technology to local cultures. This trust is the highest expression of sovereignty, because it acknowledges that power resides not in central servers or ministries, but in the collective intelligence of millions building their lives with dignity.

When the future historian looks back at this period, the measure of India's success will not be how many apps it launched, but how many communities it empowered; not how many platforms it regulated, but how many cities it trusted to design their own digital destinies.

Sovereignty will not endure through control; it will endure through confidence—confidence in our ability to build, to debate, to differ, and yet to collaborate. That confidence begins with economic freedom. A nation that is free to create wealth with purpose is strong enough to design technology with conscience.

India's digital future will depend not only on how well we regulate technology, but on how deeply we trust ourselves to create wealth with dignity. A confident economy can design confident systems. A free society can design free technologies. Only then will India's digital sovereignty be more than a metaphor—it will be a lived reality, built on the unshakeable foundation of economic freedom.

The logic of sovereignty, in every age, has followed the flow of power. In the industrial era, power was defined by those who controlled factories, capital and trade routes. In the digital era, it is defined by who controls infrastructure, innovation and attention. Beneath both lies a constant: the independence to act—to build, to decide, to err and to recover. That independence, whether for a nation or a citizen, is born of economic freedom.

For too long, our national discourse has treated economic freedom as a matter of convenience, a chapter in a budget speech or a technical debate between regulation and reform. It is far more elemental. Economic freedom is the oxygen that allows societies to imagine futures of their own making. When individuals and enterprises can create wealth without undue friction, when cities can raise resources without dependency, when governments can invest without fear of fiscal fragility, sovereignty ceases to be an aspiration and becomes a living condition.

India's challenge, therefore, is not merely to regulate technology better, but to deepen the foundations that make regulation meaningful. Digital policymaking will remain fragile until our cities, states and enterprises have the fiscal muscle and institutional authority to stand upright. This requires redistribution not of wealth, but of responsibility.

That is the real political economy of digital sovereignty. It is not only about who writes the laws or owns the platforms, but about who has the fiscal and institutional space to act without seeking permission. The Indian state must evolve from a regulatory hierarchy into an enabling architecture. Regulation has its place, but its highest purpose is to create trust—not fear—and to make room for innovation rather than merely audit its aftermath.

In this shift from control to confidence lies the essence of the next policy frontier. When citizens can see how their data is used, when innovators can trust that the rules of competition are even-handed, and when institutions are rewarded for integrity rather than obedience, economic freedom becomes the most powerful engine of sovereignty.

Sovereignty, after all, is a public good—indivisible, cumulative and self-reinforcing. Like clean air or public health, it benefits all when protected by all. It is produced not only through armies or treaties, but through the daily discipline of fairness, enterprise and trust.

The same philosophy must guide the bureaucracy. It is time to reward initiative, not inertia. The future administrator must see herself not as a custodian of files but as a builder of ecosystems—a public entrepreneur whose success is measured not by compliance but by the capacity she unlocks. The Indian Administrative Service can either remain the steel frame of yesterday or become the innovation lattice of tomorrow. Its transformation will determine whether India can convert its demographic advantage into institutional strength.

The emerging global digital order will test not just our capacity to compete, but our ability to cooperate—across cities, sectors and communities. India's task is to demonstrate that democracy can be not only a moral virtue but also an innovation advantage—that open societies, when economically empowered, can build digital systems as resilient as they are just.

If we can achieve this—align economic freedom with digital fairness, fiscal autonomy with moral purpose—then India will not merely participate in the digital century; it will help author it. And that authorship will be the truest form of sovereignty: not the power to dominate others, but the freedom to define oneself.

As a nation-state, what does sovereignty mean? There may be many operational and tactical definitions, but at an abstract level, it simply means the freedom to pursue actions that serve the interests of all people within its boundaries—not just a narrow elite. In a world of competing nation-states, no country enjoys absolute freedom; constraints are imposed externally by others and internally by its own past choices—what one may call historical karma.

Technology adds another layer of constraint. It is difficult to imagine a future with greater freedom than the present unless these constraints are first recognized, acknowledged and understood before policy is formulated. As outlined in earlier chapters, India is neither an industrial power, nor a research power, nor a financial power. These structural limitations have been compounded by past choices—an inability to make long-term risky bets, a preference for short-term outcomes, state support for locally optimized systems that do not scale globally, the absence of strategic culture, reliance on externally driven regulatory frameworks, and a limited understanding of the AI supply chain.

Yet one limitation stands above all: a poverty of ambition. It is the creeping resignation that celebrates mediocrity, accepts dependency as destiny and is content with marginal progress. We have been conditioned to believe that incrementalism is enough—that a 'whatever we get, we take' mindset is sufficient for a back-end nation.

And yet sovereignty is not an abstraction. It is personal before it is national.

It begins with the ordinary citizen knowing that their savings, their data and their choices are not at the mercy of distant corporations or foreign sanctions. If we cannot guarantee that dignity, every claim of sovereignty is hollow. Freedom without sovereignty is an illusion; sovereignty without freedom is tyranny. The challenge before us is to secure both—together, and with conviction.

Our forefathers faced a similar dilemma in the 20th century. Colonial rule was not merely political domination; it was the denial of dignity in everyday life. The freedom struggle demanded that Indians govern themselves, speak for themselves, and shape their own destiny. Later, through the articulation of the Panchsheel principles, newly independent nations asserted that freedom is indivisible—that political independence without economic and cultural sovereignty remains incomplete. The digital age brings us back to that same crossroads.

We do not deny our limitations, nor do we accept them as permanent. They can be dismantled—slowly, deliberately, and with purpose. Because if they are not, sovereignty itself loses meaning for both the people and the nation-state our ancestors created.

By refusing to acknowledge these constraints, we risk perpetuating them indefinitely. Unfortunately, public policy in India often operates detached from these realities. This tendency is evident in the domain of AI as well.

In the next chapter, we turn to the challenge of designing AI public policy—through concrete examples of what to do, and equally important, what not to do.

7

Why Public Policy Often Regulates Myths on Math but Not Math

Myths are a powerful way to describe reality because they allow the reader to focus on the essentials without getting lost in detail. Consider a popular story of a king who asked for all human knowledge to be summarized into a digestible book in the era of palm leaves. His courtiers tried repeatedly and failed; the king, in turn, repeatedly executed them. Finally, a wise man wrote three sentences on a palm leaf and sent it to the king: 'Humans are born. They live. And they die.'

One may laugh at this absurdity, but the story itself is a meta-story. The absurdity is compression, and the origin of the absurdity is the king himself, who demanded it. One may argue, therefore, that the story itself is a recursive myth—one that explains what a myth is. It is reality expressed through extreme compression, driven by a peculiar characteristic of the human mind: it craves simplicity. That craving makes it prone to severe cognitive biases. Compression satisfies the craving while simultaneously making thought numb and shallow.

Magical thinking is one such bias, and one needs only to look at human history to see how deeply it is embedded. Consider the practice of praying to gods to send rain during a parched summer or after a crop failure. It made perfect sense when we lacked knowledge of cloud formation, heat transfer, ocean currents and monsoon winds. In the absence of such knowledge, we arrived at an approximation: there exists a god of thunder and rain who can be appeased through prayer.

This approximation is as absurd as summarizing human history as 'born, live, die'. But the problem is not the absurdity itself—it is the side effects. Once we accepted that praying to gods would bring rain, there followed an increasing attempt to standardize and define the practice in exacting detail: the size of the temple, the components of the prayer, the ingredients required. Put differently, the absurdity created its own supply chain and its own class of thinkers (priests), who then developed rules to regulate this supply chain through what they considered original research, often funded by kings.

This, in turn, became public policy. It was disseminated, debated, refined through feedback and evolved through consensus.

To a modern observer, however, this entire framework would look ludicrous because it rests on foundational gaps that cannot be bridged through incremental tweaks. The system itself must be thrown out in its entirety, not improved. Yet pointing this out—that praying to gods for rain is madness—would not be well received. The supply chain built around the belief would have entrenched itself so deeply that it would become indistinguishable from the myth itself.

Studying such rain-making public policy is therefore not futile. It offers insight into a regulatory mindset that mixes reality, magical thinking, exactitude and incrementalism—even when it fails to deliver outcomes.

In this chapter, we provide two detailed examples where attempts to regulate mathematics end up resembling attempts to regulate myths—and therefore fail. We also provide a meta-analysis of why public policy often fails in emerging technology domains, along with shorter field notes on related domains and how such failures may be addressed through first-principles thinking.

Regulating E2E Encryption

Two distinct problems arise in any communication between individuals:

1. How do we keep the conversation private (or secret)?
2. How do we prove to someone else that the conversation indeed took place?

Modern surveillance technology, until recently, made it difficult to achieve the first objective at scale. Any mass communication system typically requires intermediaries to have some access to message content in order to deliver it. A postal system, for instance, could in theory read the contents of a letter, even if sealed.

Cipher and codes, or encryption, were developed to address this, allowing the sender and receiver to obfuscate messages. However, this resulted in a new problem: how do both parties agree on the encryption method? Traditionally, this was done off-band—through prior agreement on both the key and the algorithm. While workable for individual communication, this approach does not scale to population-level systems.

The solution was a shift from symmetric encryption (where the same key is used for both encryption and decryption) to asymmetric encryption (where different keys are used). Each participant generates a pair of keys: a public key, shared openly for encryption, and a private key, kept secret for decryption. This allows secure communication at scale without prior communication between parties.

However, this introduces another vulnerability: what if one were to lose one's private key? In such a case, not only future messages but also all past messages are compromised, since an adversary could have stored encrypted messages between the two parties and later decrypt them when the private key gets leaked. In technical terms, this is called 'forward secrecy'.

The Snowden revelations marked a watershed moment in the understanding of communication privacy. They exposed the extent of mass surveillance by the US (in particular) and its allies, often enabled through hidden regulations compelling intermediaries to cooperate.

This prompted the search for a new technical paradigm—one in which intermediaries are treated as potentially hostile actors. The challenge then became: how can privacy be guaranteed even when the systems facilitating communication are themselves compromised by nation-state actors?

The Signal Protocol was a result of this search, and it introduced several unique features:

1. Given the computational complexity of using (public, private) key pairs to encrypt/decrypt all conversations—as they are computationally expensive and do not work well for media such as voice, video, calls and documents—it instead uses symmetric keys (the same key for both encryption and decryption).
2. These symmetric keys are rotated (changed) for every message.
3. The key pairs that generate these symmetric keys are themselves periodically changed after every 'N' message; this process is known as the double ratchet.[103]

The net result is that conversations between parties resemble a whisper: the intermediary knows very little about what is being communicated, even though it is responsible for delivering the messages (much like a post office routing mail without reading it).

Although Signal developed this protocol, it became widely deployed when WhatsApp licensed it and rolled it out to its billion-plus users.

However, this widespread deployment also coincided with the rapid spread of misinformation and disinformation, leading to societal harms* and increasing calls for regulation. One response by Meta was to address message forwarding, which had made it easier for emotionally charged content to go viral. It introduced labels such as 'Frequently forwarded' and 'Forwarded many times' to signal virality, leaving users to exercise their own judgement.

While many considered this insufficient, the platform constraints are fundamental: the intermediary cannot access message content—including media such as voice and video. Within these constraints, limiting forwards is among the few viable interventions.

This, however, led users to ask a natural set of questions:

1. If WhatsApp does not know the content of the message, how does it know that it has been forwarded often?
2. If it knows that a message is widely forwarded, does that imply it reads messages? And if so, is it truly end-to-end encrypted?

Based on the belief that E2E encryption is merely a marketing ploy,† some proposed a more direct solution to misinformation by targeting its source. The proposal was: what if every message carried the original sender's phone number embedded within it—encrypted in such a way that only law enforcement could decrypt it upon receiving a complaint?

This concept, known as originator traceability, was advocated by law enforcement agencies,[104] academics[105] and civil society

---

*Banaji, Shakuntala, and Ram Bhat, with Anushi Agarwal, Nihal Passanha, and Mukti Sadhana Pravin, *WhatsApp Vigilantes: An Exploration of Citizen Reception and Circulation of WhatsApp Misinformation Linked to Mob Violence in India*, Department of Media and Communications, London School of Economics and Political Science, https://www.lse.ac.uk/media-and-communications/assets/documents/research/projects/WhatsApp-Misinformation-Report.pdf. Accessed on 25 May 2026.

†Meta's own actions in the past have strengthened this belief. The response of its policy team during the Free Basics movement has been widely criticized as paternalistic, with Marc Andreessen, a board member, even stating that colonialism had been good for India before; later deleting the remark without explanation following severe backlash.

actors concerned with issues such as child safety.* However, the proposal lacked technical depth,† offered little understanding of E2E encryption and was often framed as a sovereignty issue—suggesting that encryption could destabilize democracies.[106]

One of the key claims underpinning this argument was that WhatsApp stores metadata about messages on its servers for extended periods, separate from the message content, and denies doing so to avoid accountability. This claim does not hold up to scrutiny. Reverse-engineered WhatsApp clients‡ have shown that metadata—such as forward counts—is embedded within the message itself, and many features are implemented entirely on the client side by interpreting this metadata.

To illustrate, consider a system where the message inside an envelope is encrypted and inaccessible to the intermediary, but the post office writes a hidden note in invisible ink on the envelope indicating that it was sent by person X. In theory, this note can be revealed only under special circumstances—such as a law enforcement investigation using UV light—and only if the post office participates in the scheme.

However, because the underlying technology standards are open, with publicly known APIs and client libraries available in multiple programming languages, it is relatively easy to build custom clients. Unofficial clients such as GB WhatsApp—often side-loaded

*This framing—that end-to-end encryption is harmful for children because it serves as a conduit for child sexual abuse material—is what we refer to as 'pedaphrasty'. Any policy outcome can be reframed as preventing harm to children, thereby making the issue emotionally charged and consequently more difficult to contest.

†Ramesh, M., 'Whatsapp Message Tracing: Balance between Privacy and National Security Can Be Achieved, Says Kamakoti', *The Hindu BusinessLine*, 9 January 2022, https://www.thehindubusinessline.com/info-tech/social-media/whatsapp-message-tracing-balance-between-privacy-and-national-security-can-be-achieved-says-kamakoti/article62169757.ece. Accessed on 25 May 2026.

‡The popular Venom JS library is now gone from being Open Source to a commercially available client. However, enough forks exist to understand the metadata model.

to bypass restrictions like forwarding limits or auto-deletion—demonstrate this clearly.

This means that metadata itself can be manipulated. A malicious client could embed false origin information (for instance, marking a message as originating from person Y), thereby falsely implicating an innocent individual.

As Mathew Greene, a professor of cryptography, notes in his blog:[107]

> The approaches I describe above rely on a critical assumption: that all participants in the system are going to behave honestly—that is, everyone will run the official WhatsApp client, which will contain logic designed to store an escrow record on WhatsApp's servers. Nobody in this system will try to bypass this system by running an unofficial client, or by hacking their client to disable the escrow logic.

The inability of policymakers to understand technology in depth thus resulted not only in court litigation* but also in a failure to move the issue forward in the public domain. It remains stuck in a peculiar limbo. The rules mandating originator traceability are law, yet they are under judicial challenge. The courts have not issued a stay pending final judgement, and that judgement has not yet been delivered.

In the meantime, WhatsApp continues to operate. The government, however, proceeded to ban[108] other end-to-end encrypted applications used by terror groups (e.g., Threema, Element), citing the absence of a local office address in India and the lack of a designated grievance officer—a significantly lower compliance threshold. Despite these bans, app stores continue to

---

*Grouthaus, Michael, 'WhatsApp Is Suing the Indian Government over "Traceability." But What Is It?', *Fast Company*, 26 May 2021, https://www.fastcompany.com/90640728/whatsapp-is-suing-the-indian-government-over-traceability-but-what-is-it. Accessed on 25 May 2026.

host these applications,[109] and they function in India most of the time. In the rare instances where they do not, access can be easily restored through VPN services.

One possible policy response could have been to explore a balance between national security and individual privacy by investing in research and commercial alternatives that preserve end-to-end encryption while addressing legitimate state concerns. Yet there have been no meaningful initiatives—either commercial or academic—that demonstrate such a balance.* The Army's signal intelligence unit did roll out an 'army-only' application claiming E2E encryption, but its design is opaque, and it cannot be independently audited or reverse-engineered as it is not available in the public domain.

To summarize:

- The state chose to intervene in a deeply technical domain (E2E encryption) through legislation that diverged so far from the realities of the technical domain that it amounts to regulating a myth about mathematics rather than mathematics itself.
- The intervention failed spectacularly and could not even take down some of the banned apps from the app stores; even if it did, it would not have made any difference. App stores are only one distribution channel. APK files can circulate through websites, file-sharing links or even messaging apps such as WhatsApp, making such bans easy to bypass.
- Attempts to build alternatives were half-hearted, resulting in systems that lack both credibility and functionality, thereby weakening the state's position despite its self-perception as a digital superpower.

---

*Zoho Corporation's Arratai explicitly does not support E2E encryption. Other apps, such as Troop Messenger and Namaste Bharat, make such claims, but none of them is auditable, as they neither publish source code nor indicate the cryptographic approach used through white papers, technical blogs or other means.

Sovereignty—the capacity to act in pursuit of one's interests—is not merely about drafting rules (such as GDPR), but about possessing the capability to implement them in the real world and manage their consequences. This requires technical depth, an understanding of domain constraints, sustained innovation and long-term execution. In the absence of these, public policy on E2E encryption risks becoming an exercise in myth-making—about both technology and the state itself.

Freeing AI from Anthropomorphizing

Just as societies once created rain gods to explain and influence natural phenomena, we now respond to the unfamiliar nature of predictive systems by ascribing personality to them. When we say 'the AI thinks', 'it refuses to answer' or 'it believes', we attribute human qualities to what are, in reality, statistical weights and vector embeddings. In doing so, we begin to treat these systems as if they possess will, intent or conscience.

This habit is increasingly visible in corporate boardrooms, policy speeches and regulatory frameworks. And therein lies the danger. The moment we anthropomorphize systems that are fundamentally probability distributions over possible outputs, we invite flawed governance, misplaced accountability and a distorted sense of trust.

An AI model—whether used for credit scoring, fraud detection or conversational interfaces—does not 'decide' in any human sense. It computes outputs conditioned on its training data, architecture and inputs. When it denies a loan or flags a transaction, it is not exercising judgement; it is executing statistical inference. Its outputs are shaped by probabilities, biases in training data and model design—not by empathy, intent or malice.

This distinction is particularly critical in the Indian financial context. Regulatory frameworks, including the RBI's recent committee report on FREE-AI,* often implicitly treat AI as a quasi-human agent—something that can be 'trustworthy' like a banker or 'transparent' like a board.

But AI is neither. It is a probability machine. Misunderstand its nature, and regulation will target its shadow rather than its substance.

Financial systems already operate with probabilistic models. No one asks the Black–Scholes formula what it 'thinks' about volatility; its assumptions are examined, its parameters stress-tested and its outputs validated. A value-at-risk model is not 'confident' or 'anxious'; it produces a distribution of expected losses that expands or contracts with inputs. These disciplines—model validation, back-testing, independent review—are well established in model risk management.

AI governance must extend these principles, not replace them.

We do not teach probability to be moral; we teach it to be precise. AI is no different. Anthropomorphizing probability distributions distorts policy in three fundamental ways:

- First, it encourages the belief that bias or error can be 'trained out' by appealing to values, as one might coach an employee. Bias is a statistical property of data and architecture. Mitigating it requires data audits, re-weighting, constraint optimization and continuous post-deployment monitoring.
- Second, it tempts supervisors to rely on qualitative impressions—'is the AI behaving fairly?'—instead of measurable thresholds—'does the model's false negative rate differ by more than X per cent across protected classes?'
- Third, it fosters a false sense of relational trust. Customers and staff may imagine they are interacting with an

*Reserve Bank of India, *FREE-AI Committee Report: Framework for Responsible and Ethical Enablement of Artificial Intelligence*, https://rbidocs.rbi.org.in/rdocs/PublicationReport/Pdfs/FREEAIR130820250A24FF2D4578453F824C72ED9F5D5851.PDF. Accessed on 25 May 2026.

intelligent counterpart; in reality, they are dealing with a brittle calculator that fails not at scale, but at the margins—and often silently.

When we describe AI as 'responsible' or 'ethical', we are not identifying a genuine moral capacity; we are projecting human traits onto a system that is, at its core, a probability distribution generating outputs from learnt patterns. A model has no intentions, values or conscience. Any appearance of ethical behaviour is the result of human design and curation—training it to align with our constraints. Responsibility and ethics reside entirely with the humans who design, deploy and regulate these systems, because only they possess moral agency. To forget this is to blur the line between tool and actor, and to build policy for an illusion.

In India's digital financial infrastructure, these statistical realities are not theoretical. Systems such as UPI, the Account Aggregator framework and Aadhaar-linked services operate at high volume and low latency. A miscalibrated fraud detection model can freeze thousands of accounts within seconds. A drift-affected credit scoring model—trained on data from a different temporal context—can systematically misjudge borrowers in the post-DPI era because input distributions have shifted. This is not a 'change of mind'; it is a covariate shift problem.

Fraud and AML systems offer another caution. These are essentially anomaly detectors operating over transaction graphs. False positives can overwhelm back-office systems; false negatives can miss coordinated attacks. Supervisory thinking must therefore focus on Type I and Type II error trade-offs, detection thresholds and tail-risk exposure—not narratives about 'AI vigilance'.

A regulator who understands the statistical nature of AI would frame policy differently. Explainability would not mean asking 'what the AI was thinking', but documenting chains of data transformation, feature importance and performance metrics.

Accountability would not involve holding 'the AI' responsible, but ensuring that human operators can trace, test and, if necessary, deploy kill switches. Risk classification would be driven by variance, drift rates and uncertainty under stress scenarios—much like macroprudential stress tests in finance.

This shift also reshapes consumer protection. If a customer is denied a loan, the explanation must refer to the statistical drivers behind the score—not anthropomorphic metaphors about 'judgement'. If a model behaves erratically in volatile markets, the relevant question is not 'why did it change its mind?' but 'what shifted in the input distribution, how was it detected, and how quickly did the institution respond?' Confidence intervals, not narratives, are what matter.

Embedding this probability-literate mindset requires rules that mandate measurable performance bounds, bias metrics, drift thresholds and scenario-based robustness testing. It requires model cards, audit trails and stress-test documentation to be as non-negotiable as KYC or capital adequacy norms. It also means recognizing systemic correlation risk: if multiple institutions rely on the same third-party model, their errors become correlated—amplifying systemic shocks, much like identical credit-risk models did before 2008.

The RBI's report gets much right—its emphasis on governance, data controls and systemic risk is both overdue and necessary—but it falters when it frames 'responsible' and 'ethical' as if these are attributes the machine itself can embody. Can a probability distribution 'be fair' in a moral sense, or does it merely satisfy a statistical definition of parity? Can an algorithm 'act responsibly' when it has no agency—only parameters and loss functions? Who is being trusted—the model, or the humans behind it? And if ethics is reduced to a comforting narrative about AI 'doing the right thing', how do we detect when silent data drift turns the right outcome into the wrong one? By anthropomorphizing, do we risk the same complacency that comes from trusting a polished risk model without interrogating its assumptions?

India's regulators are still early enough in their AI journey to avoid the Western trap of 'trust frameworks' built for imaginary digital actors. We can instead regulate AI for what it is—statistics in motion—placing accountability firmly on human actors and measuring outcomes in probabilities, not personalities.

We do not need to strip poetry from our conversations about technology. But we must strip it from policy. The sooner we stop humanizing algorithms, the sooner we can regulate them effectively—as systems that shape credit flows, fraud detection and financial stability. In a country where a false negative can mean exclusion and a false positive can mean insolvency, this is not a semantic distinction—it is the foundation of sound, responsible governance.

Generative AI introduces additional challenges even at a statistical level—challenges that existing regulatory frameworks do not yet fully grasp, regardless of whether the model is large or small. Consider datasets in Indian languages. For many languages, they are sparse or non-existent because educational and research institutions did not systematically preserve cultural and linguistic artefacts. Moreover, formal language learning often ends at the school level.

Dataset creation, curation and labelling are already several times higher*—estimated at 3x to 17x the cost of training—even for English. For local languages, where data is sparse, the cost is likely to be at least twice as high.

As Halvar Flake observes:[110]

> Alignment and safety for LLMs mean that we should be able to quantify and bound the probability with which certain undesirable sequences are generated. The trouble is that we

*Zhu, Yuxuan, and Daniel Kang, 'Human Data is (Probably) More Expensive Than Compute for Training Frontier LLMs', *Daniel's Substack*, 11 August 2025, https://ddkang.substack.com/p/human-data-is-probably-more-expensive. Accessed on 25 May 2026.

largely fail at describing 'undesirable' except by example, which makes calculating bounds difficult.

Principle	*Analysis*
Trust is the Foundation	Conveniently ignores the power asymmetry that creates distrust and fails to analyse the impact of AI deployment within this context.
People Safety	Treating AI as an anthropomorphic entity with independent agency is equivalent to treating rain as a consequence of appeasing a rain god.
Innovation over Restraint	Largely accurate in its approach to emerging technologies.
Fairness and Equity (e.g., AI should uphold fairness)	Treats AI as a human entity, while fairness itself remains undefined and unquantified.
Accountability	Accurate, as it correctly assigns agency to those who deploy AI, not to AI itself.
Understandable by Design	Conflates different types of AI (predictive vs generative). While predictive systems can be partially understood using classical statistical techniques, transformer-based models are not yet explainable even by their creators. Moreover, lack of explainability (e.g., in credit denial) is already permitted as a business decision under existing regulations.
Safety, Resilience and Sustainability	Energy efficiency is problematic: transformer models require GPUs for attention mechanisms, and GPU clusters demand exponentially increasing power for training and inference.

When it is not possible to specify a bound within which an LLM will generate undesirable sequences—even in languages with rich labelled datasets—the problem becomes untenable in sparse local language contexts. Additional constraints emerge:

1. How will regulated entities monetize AI projects when the average income of 90 per cent of the population is only ₹1.3 lakh* per month, even assuming not all are local language users?
2. Given well-documented power asymmetries† between government, regulators, banks and consumers—leading to entrenched distrust—what does an 'AI-driven system reinforcing trust'‡ even mean?
3. If inference costs trend towards the cost of electricity while training costs remain orders of magnitude higher, what does 'energy efficiency' practically imply?

The table above summarizes our analysis of the misalignment of the seven principles with both technological reality and the financial sector's local context.

Leaving aside principle-level misalignment, the report argues for 'Equitable access to high-quality data', noting that: 'Data availability is asymmetric, i.e., large incumbents have access to huge datasets that smaller REs lack. It is often stored in non-standard formats, making it difficult to use. As a result, substantial time and effort must be spent collecting, cleaning, and transforming data before it can be used in AI.'[111]

---

*'Top 1% Indians' Share in National Income Is Higher Now than under British Rule: Data', *The Hindu*, 23 March 2024, https://www.thehindu.com/data/top-1-percent-indians-income-share-is-higher-now-than-under-british-rule-data-point/article67979494.ece. Accessed on 25 May 2026.

†See Chapter 3.

‡Reserve Bank of India, *FREE-AI Committee Report: Framework for Responsible and Ethical Enablement of Artificial Intelligence*, https://rbidocs.rbi.org.in/rdocs/PublicationReport/Pdfs/FREEAIR130820250A24FF2D4578453F824C72ED9F5D5851.PDF. Accessed on 25 May 2026.

This framing is deeply problematic. It conflates financial transactional data with the high-quality labelled datasets required for training LLMs—two fundamentally different categories of data. Financial institutions primarily hold structured data on loans and repayments, which is already accessible via credit bureaus for a fee.

By proposing a centralized data lake and mandating that models built on such data be open-sourced, the report effectively advocates forced socialization through a central directive, while framing it as democratization to benefit smaller players.

This reflects the same 'infant industries need protection' mindset that has historically shaped India's industry policy—despite repeated evidence of its failure to produce globally competitive firms and breakthrough innovation. Extending this approach to regulated financial entities, which are based out of India and require an explicit license from the RBI to even operate, compounds the distortion.

In substance, the proposal resembles earlier attempts to create repositories of non-personal data[112] (NPD) for universal access. These discussions continue across legal, advocacy and industry circles without formal policy adoption. Concerns around transparency, misuse (including de-anonymization) and data monopolization remain unresolved, and legislative focus has largely shifted to personal data protection.

The report is also not based on operational reality when it claims that it draws 'from earlier examples, such as the introduction of BSBDA Small accounts with simplified KYC'[113]—despite well-documented structural issues with KYC, discussed extensively in earlier chapters.

Further, the proposed funding—₹5,000 crore as initial seed capital and ₹1,000 crore annually—fails to account for the capital intensity and scaling dynamics of the AI industry.* It ignores the

*This is lower than even the allocation of a city region in China. For instance, Shenzen alone put a target of ¥10 billion fund + ¥4.5B raised with an annual ¥950M in targeted support.

underlying physics of compute, data and model development.

Finally, the recommendation for full transparency into AI operations—including detailed bills of materials, board-level oversight committees, regulatory audits and compliance frameworks managed by self-regulatory organizations—creates an overwhelming compliance burden. Regulated entities seeking guidance on AI deployment may find themselves navigating requirements such as 'AI for affirmative action' without clarity on implementation or business viability.

To summarize the broader critique: public policy in India, particularly in emerging technologies, tends to operate through a 'status-plus-plus' and 'state-plus' incrementalism. It misreads both the trajectory and the foundational nature of the technology, then attempts to force-fit it into existing regulatory frameworks. This force-fitting is institutionalized through audits, compliance mandates and new organizational structures (e.g., national AI safety institutes), ultimately constraining innovation and embedding mediocrity.

This pattern recurs across sectors, and emerging technologies are no exception.

'Yathā dṛṣṭiḥ tathā sṛṣṭiḥ'—
As our perception, so the world we create.

Across centuries, India has understood that *nīti* (policy) and *dharma* (ethics) must remain aligned, or power turns blind. Yet in the digital age, governance often confuses faith in code with understanding of consequences. Like ancient myths that explained monsoons through the moods of gods, modern policy risks mistaking slogans for science and intent for infrastructure.

These field notes trace that enduring tension between myth and mathematics—between narrative and number, between how systems are imagined and how they actually behave. Each vignette reflects what happens when the poetry of governance outruns its precision, and when sovereignty must be reclaimed not only in law but in logic.

Policy, like mythology, is a story we tell ourselves about control—about how rules, code and numbers can contain the forces of change. Yet every era discovers that its equations are shaped as much by fear and belief as by reason. These fragments explore that uneasy boundary between myth and math—where public imagination meets technical reality, where good intentions falter under hidden assumptions, and where sovereignty is ultimately determined not by declarations but by design.

The Whisper Networks of Hong Kong

When the streets of Hong Kong were filled with umbrellas and chants, a quieter revolution unfolded in the palms of people's hands. Protesters abandoned mainstream apps and flocked to Signal and Telegram—not out of ideology, but instinct for survival. Policy, meanwhile, continued to speak the language of oversight and traceability. What governments see as opacity, citizens experience as safety. In those encrypted whispers lay a simple truth: when trust in institutions fades, people turn to math. The policy lesson endures—regulate behaviour, not belief; reality, not myth.

When an Algorithm Took the Stand

In a Wisconsin courtroom, a judge once read aloud a risk score from an algorithm called COMPAS. The machine declared a defendant 'high-risk'. Years later, journalists revealed that its errors mapped neatly onto race. The algorithm hadn't discriminated consciously; it had simply reproduced the patterns embedded in its data. But the public heard it differently—as if the algorithm had lied, as if it possessed intent. In that moment, a probability model became a moral villain. Regulators demanded ethics frameworks yet ignored calibration charts. Society wrote myths around software instead of reading its statistics. That's how myths replace math.

The Ghost in the Newsfeed

When the Cambridge Analytica scandal broke, it wasn't a hack that terrified the world—it was the realization that persuasion itself had been privatized. Millions were profiled, their fears and hopes recycled into targeted political messaging. Regulators rushed to write privacy laws, yet missed the deeper question: who owns the algorithmic lens that shapes perception? The real breach wasn't of data; it was of democracy's collective mind. We built myths around 'consent', while the math of influence quietly rewrote our sense of truth.

The Day the Cloud Went Dark

One ordinary afternoon, a single Amazon Web Services region crashed—and with it, thousands of hospitals, government websites and logistics systems around the world. The outage lasted hours, but it stripped bare an illusion that had endured for decades. Nations had mistaken service-level agreements for sovereignty. The math of dependency was suddenly visible: a handful of private infrastructures powering public life. No parliament had voted for this fragility; it had accumulated silently, one convenience at a time. The myth of control met the math of concentration—and lost.

But how does one move from diagnosing faults and failures to laying out how public policy should actually work? We advocate a first-principles approach, with specific methods detailed below.

Map Intention from Invention

In the modern contest for power, the battle is as much for narrative as it is for market share or technological supremacy. Technologies today do not merely disrupt industries or reshape markets; they shape the stories a nation tells about modernity, agency and progress. Those who articulate these stories first often set the norms that others must follow. This is why the framing of emerging technologies matters as deeply as their deployment, and why India must move from adopting external frameworks to authoring its own.

When artificial intelligence is presented narrowly as a tool for efficiency or surveillance, the debate shrinks to questions of productivity and security. But when it is framed as a faculty that mirrors—and may even govern—human decision-making, the conversation expands to questions of agency and dignity. Similarly, when biotechnology is described as an export industry, policy centres on incentives; when it is seen as a force that could reshape health or heredity, deeper ethical and sovereign concerns emerge. The narrative determines not only what is regulated, but why it is regulated at all.

In a country as vast and layered as India, the stakes of these narratives multiply. Policy cannot be written solely from the vantage point of metropolitan centres or industry conferences. Every regulatory choice creates second- and third-order effects, often far removed from where the law is drafted. A data protection policy designed for urban consumers may inadvertently raise compliance burdens for rural innovators. A framework for online speech moderation shaped by national debates can become, in practice, a quiet erasure of regional cultures. At this scale, unintended consequences can rival intended outcomes.

Understanding these consequences is not a technocratic detail; it is an act of strategic foresight. In the emerging architecture of global power, sovereignty is no longer defined only by territory or economic output. It is increasingly shaped by who controls data flows, algorithmic architectures and the default settings of digital platforms. These systems—often designed abroad—now influence domestic commerce, public discourse and even electoral trust. To remain sovereign in this environment is to have the capacity to question the values embedded within them—and to shape or contest them when necessary.

Yet sovereignty at scale presents a uniquely Indian challenge. Policies must be ambitious enough to set national direction, yet flexible enough to reflect local realities. In a country where a single rule can affect hundreds of millions overnight, governance must extend beyond the average user to those at the margins. This challenge is compounded by the speed of networked technologies, which outpace traditional consultation cycles. Strategic policy must therefore balance reach with nuance and speed with deliberation.

This demands a different habit of mind within policy circles: the discipline to treat narrative as a strategic lever, not a communications afterthought; the patience to map consequences beyond immediate stakeholders; and the humility to recognize that sovereignty now resides as much in invisible architectures of code and standards as in visible laws.

To do otherwise is to fight yesterday's battles—to regulate tools rather than question the purposes that drive them, to draft frameworks reactive to headlines rather than proactive in vision, and to protect markets without protecting the values that underpin them. To do it well, however, is to ensure that India does not remain merely a user of technologies authored elsewhere, but becomes an author of its own digital and scientific destiny.

In that lies the true task of governance: aligning what is technically possible with what is nationally desirable—and doing so in a voice unmistakably our own.

In policy circles, there is often an unspoken assumption that understanding the mechanics of an emerging technology suffices to shape its trajectory. Expert committees convene, white papers are commissioned, and strategy notes circulate across ministries and think tanks. Yet beneath this surface lies a deeper truth, rarely acknowledged: the roots of any transformative technology extend far beyond its technical specifications or immediate market applications. They draw from a primal human impulse—to overcome the constraints of nature, to bend uncertainty to our will, and to expand the boundaries of what it means to be sovereign and human.

Artificial intelligence, so often reduced to algorithms and data pipelines, is not merely an efficiency tool or an instrument of surveillance. At its core, it is humanity's audacious attempt to externalize cognition itself—to replicate, at scale, a faculty once thought to belong exclusively to conscious minds. This is not merely a technical project; it is a civilizational wager. Those who shape AI systems are, in effect, deciding what forms of reasoning and bias the future will inherit. They write the silent compression functions that determine which truths are amplified and which are forgotten. If policy discourse remains preoccupied with deployment and risk frameworks while ignoring this deeper philosophical terrain, we end up three-fourths blind—engaging with the architecture while neglecting the foundation.

Another industry. Another example. Yet the same underlying impulse.

Biotechnology and genetic engineering present a similarly profound challenge. In policy rooms, they appear as tools to eradicate disease, increase agricultural yields or enhance national competitiveness. Yet the deeper force driving genetic research is far older: the human refusal to accept biological destiny—ultimately, a refusal to accept death itself. It is about claiming agency over heredity and lineage, reshaping them according to human intent. For a nation, this becomes a question of demographic sovereignty: whose norms will decide which traits are edited, which therapies are prioritized, and who gains access to life-extending science? Without engaging with this underlying drive, regulatory debates risk degenerating into compliance exercises, blind to their moral and geopolitical consequences.

Consider also the revolution under way in quantum technologies. It is tempting to view them through the lens of faster computation or next-generation encryption and decryption. But the deeper reality is that quantum research challenges the very architecture of reality itself. It offers the possibility of computing the previously incalculable and modelling the previously unknowable. The implications extend far beyond markets—to national security doctrines, global financial systems and the delicate balance of deterrence. In this domain, failing to grasp the philosophical roots of the technology is to misunderstand why states invest in it with such intensity and secrecy.

India's renewed focus on space technology similarly transcends the stated goals of satellite connectivity or remote sensing. The underlying imperative is to secure a presence in new domains of sovereignty—to ensure that when humanity expands beyond Earth, India arrives not as a passenger but as a co-architect of that frontier. It is about strategic depth not only in physical space but in narrative power. Those who shape the norms of space governance—resource extraction, orbital security and habitation—will wield disproportionate influence in the century to come.

The same logic applies to energy technologies, from fusion research to advanced battery storage. These are often framed as responses to climate obligations or growth targets. Yet at a deeper level, they represent an attempt to decouple human progress from finite planetary constraints—to transcend scarcity itself. Nations that master these technologies do not merely gain new industries; they redraw the geopolitical map by redefining the relationship between growth and limitation.

To meet these challenges, Indian public policy must move beyond the comfort of technical briefings, regulatory sandboxing and reactive legislation. It must begin with first principles. Why does a particular technology emerge at this moment in history? Whose anxieties and ambitions gave rise to it? Which global narratives does it reinforce or challenge? What will it do to the texture of Indian society—not merely in terms of aggregate GDP, but in the lived realities of caste, gender and the urban–rural divide? And crucially, what does it mean for India's pursuit of strategic and digital sovereignty in a world where code, data and biology are becoming the new loci of power?

Such reflection is not a philosophical indulgence; it is a precondition for policy that is anticipatory rather than reactive, and for governance that aligns national interest with ethical stewardship. The technological forces shaping this century are not neutral tools; they are extensions of human will, saturated with the hopes, fears and biases of their creators. If we fail to recognize this, our policy responses will remain shallow—technically informed, yet existentially incomplete—seeing only one-fourth of the world while leaving the remaining three-fourths unexamined.

For a country of India's scale and civilizational depth, the task is not merely to regulate what others invent, but to shape why and how such inventions emerge—and to determine the values embedded within them. Only then can sovereignty be meaningfully claimed, not merely in markets and institutions, but in the ideas and ideals that define India's contribution to the next global epoch.

Understanding FOMO as a Foundational Epistemic Risk

In the clamour of new acronyms, policy roundtables and regulatory drafts, it is easy to forget a first principle: technology is never neutral. It carries the philosophy of its makers, the incentives of its financiers and the cultural assumptions of its architects. In a country as large, diverse and ambitious as India, public policy cannot simply trail behind innovation—cleaning up excesses and drafting rules after crises have already unfolded.

This is not a call for scepticism alone. Rather, it is an invitation to pursue deeper understanding. Consider the technologies reshaping our daily lives today: artificial intelligence, genomic editing, quantum computing, extended reality and decentralized finance. Each promises productivity, speed and convenience. Yet beneath these promises lie more fundamental, and often uncomfortable, questions about identity, agency, fairness and sovereignty. If policymakers restrict their gaze to operational risks or compliance mechanisms, they risk mistaking the noise of regulation for the substance of governance.

Behind many policy initiatives—even those framed in the language of boldness and national interest—lies a hidden driver rarely acknowledged in formal discourse: the fear of missing out. In closed-door meetings, seasoned bureaucrats and policy advisors sometimes admit, if only half in jest, that what truly moves files forward is not always conviction, but the anxiety that another nation may regulate faster, claim the narrative or set the standards first. This instinct is not unique to India; it is present in every polity navigating rapid technological change. But it is particularly potent in large democracies, where accountability is diffuse and public expectations are high.

In these moments, a subtle but consequential shift occurs—from asking, 'What does India genuinely need?' to asking, 'What are others doing, and how do we avoid being left behind?' This question, reasonable on the surface, becomes problematic when it bypasses a deeper analysis of the domestic context. Technologies born in Silicon Valley or Shenzhen carry with them embedded social assumptions, economic models and cultural values that do not translate seamlessly into Indian realities. The pressure to announce a comparable framework, sandbox or regulation often overtakes the slower, more uncomfortable work of first-principles thinking: examining who will be most affected, what unintended markets may emerge, and how existing social hierarchies may be reinforced or disrupted.

At its root, this policy FOMO arises from two sources. First, the belief—often justified, but not iron-clad—that global standards, once established, become difficult to challenge later. Second is the political calculus that it is safer to be seen to act—even reactively—than to risk appearing indifferent to disruption. This bias towards action offers short-term protection from criticism. But in the longer run, it risks locking the nation into frameworks poorly suited to its scale, diversity and developmental priorities—an example being the adoption of sanctions-driven FATF-inspired KYC norms without adequate contextual adaptation.

The pattern repeats across sectors. In data governance, India's early frameworks mirrored elements of European regulation without always accounting for domestic infrastructure gaps. In fintech, regulatory enthusiasm was shaped as much by global narratives as by a granular understanding of credit behaviour in small towns and rural regions. Even in the race to announce AI strategies, there has often been greater emphasis on producing documents than on debating what such strategies should meaningfully protect, enable or restrain within the Indian context.

What remains underappreciated is that the network effects of regulation can be as consequential as those of the technologies they seek to govern. A hastily drafted rule can concentrate power among a handful of incumbents who alone can bear the burden of compliance. It can deter domestic innovation while leaving global players relatively unaffected, or redirect capital towards speculative applications rather than socially meaningful ones. Once such dynamics take hold, they become difficult to reverse.

It is tempting—especially for conscientious officials under public scrutiny—to equate speed with wisdom. Yet the deeper responsibility of governance is to slow the moment just enough to see beyond immediate comparisons with other nations. The central question should never be whether India has a framework as quickly as others, but whether the framework emerges from an honest accounting of our unique demographic realities, economic structural constraints, vulnerabilities and domestic priorities.

Policy, at its best, is not an exercise in imitation but a craft of adaptation and anticipation. It must have the courage to ask not only what a regulation achieves on paper, but what it triggers in behaviour, markets and power structures not yet visible. It must resist the fear of being left behind long enough to understand where the path ahead actually leads. For in the governance of technology, the greatest risk is not delay alone, but the quiet certainty that emerges when urgency replaces understanding.

Take artificial intelligence as a starting point. At first glance, it appears to be an engineering challenge of models, data and algorithms. Yet its true power—and its risk—lies in how it reshapes social hierarchies, redistributes authority and potentially erodes the role of human judgement. An AI system that selects candidates, assesses creditworthiness or generates judicial summaries is not merely a more efficient tool. It embeds assumptions, biases and values that may conflict with constitutional principles or societal expectations. To regulate AI effectively, policymakers must confront a more difficult question: in a society as plural and contested as India, what does it mean to automate decision-making? What must remain irreducibly human?

The same depth of inquiry applies to data sovereignty. Too often, the debate is reduced to server location or data processing rights. Yet the more fundamental issue is one of power: who exercises control when a nation's data flows through global infrastructures? Does sovereignty imply isolation, or does it require shaping global norms so that Indian citizens' data cannot be monetized, weaponized or manipulated? Without clarity on this question, policy risks becoming either defensive or symbolic—building walls where bridges are needed, or opening gateways without sufficient safeguards.

Consider blockchain and decentralized finance. Beyond concerns of volatility and fraud lies a deeper question: does disintermediation genuinely empower individuals, or does it simply shift power into less visible forms—embedded in protocols and governance structures? Similarly, quantum computing is not only about cryptographic disruption; it is about which nations control the next leap in computational capability, and how that control reshapes strategic balances in trade, defence and influence. These are not technical issues to be delegated to committees; they demand the imagination of statesmanship, not merely the caution of regulation.

The challenge is not unique to India, but it is more acute given our scale and complexity. Few democracies must simultaneously navigate a billion-plus population, linguistic diversity, entrenched inequalities, and the competing demands of rapid growth and social justice. In such a context, borrowed regulatory playbooks are insufficient. Policymakers must cultivate deeper capabilities—technologists who think philosophically, philosophers who understand systems, and economists who appreciate technological dynamics. Governance must extend beyond rule-making; it must become an exercise in national imagination.

Why does this matter so urgently? Because emerging technologies are not merely transforming markets; they are redefining sovereignty itself. Where nations once secured borders with armies, influence now flows through digital platforms, cloud infrastructure, standards bodies and data regimes. If India fails to ask the deeper questions now, it risks becoming a rule-taker rather than a rule-maker—its citizens reduced to data points within systems designed elsewhere, its enterprises compliant but strategically dependent.

True strategic autonomy in the digital age requires foresight—not merely in the sense of predicting next quarter's innovation, but in cultivating a 50-year view. It demands recognition that certain capabilities—domestic chip design, open-source AI models, indigenous cryptography, multilingual NLP—are not optional add-ons; they are the infrastructure of sovereignty itself. It requires seeing technology not as an industry to be supervised, but as a societal force to be shaped in alignment with constitutional values and national priorities.

This is why India's public policy must move beyond the comfort of reactive rulemaking. It must confront existential questions directly: what do we wish to preserve as humans in an age of automation? How do we ensure that data generated in India builds Indian capabilities? What role must the state play in enabling strategic innovation that markets alone will underinvest in? And crucially, how do we align technological progress with dignity?

These are not abstract philosophical concerns. They mark the difference between becoming a digital colony and a digital civilization—between policies that chase headlines and those that shape history. They cannot be resolved by importing frameworks or mimicking what appears to work elsewhere. They demand a policy culture that is comfortable with ambiguity, attentive to first- and second-order effects, and anchored in the lived realities of Indian society.

If there is a single message for Indian policymakers confronting the next wave of emerging technologies, it is this: resist the seduction of regulation as an end in itself. Return instead to the deeper questions of purpose, power and impact. For in technology, as in governance, what we choose to ask often determines what we dare to build.

History reminds us that the origins of technology are rarely purely technical. They are rooted in geopolitical anxieties, commercial ambitions and civilizational impulses—forces that often remain invisible until much later. When the British built railways across India, the official justification lay in economic modernization and administrative efficiency. Yet the underlying intent was to reinforce control over the subcontinent, enabling rapid troop movement and resource extraction. The technology itself was neutral; its deployment was not. And its imprint on the Indian landscape endures.

In the decades after independence, India's push into atomic energy was driven not only by developmental aspiration but by a profound recognition of vulnerability. Leaders such as Homi Bhabha and Vikram Sarabhai saw scientific capability as a bulwark against strategic dependence. They understood that sovereignty in a post-colonial world could not rest solely on territory or manpower; it required mastery over knowledge systems that others could neither easily supply nor withhold. This was as much a philosophical stance as it was a policy choice.

Yet even in these efforts, policy sometimes underestimated broader social consequences. The Green Revolution, while indispensable for food security, brought ecological strain, regional disparity and new hierarchies of wealth and caste. What began as a scientific intervention evolved into a force that reshaped rural power structures for generations. These are not arguments against progress, but reminders that technology never arrives alone. It brings with it complex shadows, and wise policy must learn to see within those shadows what is obscured by the glare of immediate benefit.

In more recent times, digital identity systems and data governance debates have revealed similar blind spots. Designed for inclusion and efficiency, large-scale digital infrastructures have at times deepened exclusion by failing to account for the lived realities of the poor, the undocumented and the marginalized.

Anonymized vignettes from internal policy deliberations reveal a recurring pattern: senior officials often acknowledge that 'ground reality may differ', yet decisions are driven by metrics, timelines and targets rather than by questioning the foundational assumptions of the technology—even when contradictory evidence is available. Within regulatory bodies, younger officers quietly admit that they are encouraged to treat emerging technologies as discrete policy verticals rather than as cross-cutting societal forces. It is easier to produce a strategy paper on AI for agriculture than to interrogate what AI implies for human agency in rural economies. Yet it is precisely these foundational questions that must precede any meaningful regulatory framework.

Strategic thought, as argued by scholars from Kautilya to Morgenthau, begins not with tools but with intent. The *Arthashastra* reminds us that sovereignty is secured not merely through power, but through understanding the sources and limits of power in a changing world. Morgenthau, writing in the age of nuclear weapons, warned that technology amplifies power asymmetrically, and that international order cannot ignore these asymmetries without consequence. These insights remain deeply relevant as we confront technologies capable of reshaping biology, cognition and computation itself.

There is a persistent temptation in policy circles to respond to complexity by fragmentation: AI is treated as digital, biotechnology as health, quantum as science and technology. Yet this compartmentalization obscures how these technologies converge—often amplifying each other's disruptive potential. AI accelerates drug discovery; biotechnology generates data that feeds AI systems; quantum computing promises breakthroughs across both domains. The real strategic question is not how to regulate each silo, but how to anticipate the systemic transformations emerging from their intersection. This is not an abstract concern. China's approach—treating modern technologies as an interconnected sequence of levers[114]—has enabled it to achieve breakthroughs across domains within two decades.

We must also recognize that the global contest is not merely for markets or patents, but for narrative supremacy. Those who define the purpose of technology often end up setting its norms. For centuries, the West has shaped the philosophical and ethical vocabulary of science—from Enlightenment notions of progress to contemporary debates on digital rights. If India remains three-fourths engaged only at the level of technical deployment, it risks ceding the deeper contest over meaning and legitimacy.

We are living through a rare historical moment where science, markets and geopolitics are simultaneously reconfiguring the frontiers of power. Whether in code, data, biology or orbital space, the contest is not only about ownership of assets but about defining why those assets matter. Policy must therefore evolve into a discipline of epistemic thinking—one that speaks not only in the language of compliance and enablement, but also in the deeper vocabulary of purpose and consequence.

In this context, the fear of missing out ceases to be a mere bureaucratic reflex; it becomes an epistemic risk—far harder to correct once embedded. By chasing symmetry with the regulatory moves of others, we risk neglecting the first duty of policy in a democracy of India's scale: to ask why a technology matters here, in this society, at this moment.

In every age, policymaking has meant navigating the frontier between the known and the unknown. But in this century, that frontier is neither stable nor distant. It is immediate, fluid and often violently unpredictable. The technologies reshaping markets, societies and even human identity—from artificial intelligence to genetic editing, from quantum computation to synthetic biology—do not evolve in linear increments. They advance in bursts: sudden leaps, unexpected convergences and, at times, abrupt reversals.

To respond to such a landscape with policy frameworks conceived as fixed and final is to risk irrelevance by design. The very character of these technologies defies static governance. Their capacity to learn, scale and recombine ensures that their trajectory cannot be fully foreseen at the point of regulation. What emerges, therefore, is not a call to abandon structure or discipline, but to reconceive policy as something closer to a living architecture: rigorous in purpose, yet adaptive in form; capable of revision not as an admission of failure, but as a sign of institutional maturity.

In practice, this requires policymaking to evolve into a process that resembles an iterative system. Just as complex software is maintained through updates, patches and version upgrades, policy too must adapt as evidence deepens, contexts shift and unintended consequences surface. This is not an argument for volatility in governance, but for resilience rooted in openness to correction.

Yet adaptability alone is insufficient. It must be grounded in inclusivity and balance. The effects of emerging technologies are rarely evenly distributed. They amplify some voices while muting others, open new markets while entrenching old inequalities, and create value that is often captured far from the communities most affected. To design policy without incorporating these perspectives is to govern with partial sight—mistaking the view from the centre for the view of the whole.

Inclusivity, in this sense, is not token consultation or procedural formality. It is the recognition that no single regulator, ministry or expert group—however capable—can fully grasp technologies that cut across sectors, demographics and social layers. A balanced process actively seeks dissent, tests assumptions against lived realities, and remains alert to how regulation itself can generate new network effects, shifting power in ways that may later prove difficult to reverse.

There remains, within many institutions, a belief that stability is best preserved through infrequent and incremental change. Yet in an era where disruption is the norm, true stability emerges not from rigidity but from continuous recalibration. A regulatory framework that fails to adapt becomes, in effect, a vulnerability—exploited by those quickest to identify its blind spots, or rendered obsolete by the very innovation it seeks to guide.

This does not diminish the need for clarity of purpose; it heightens it. If policy is to function as an evolving system, it must remain anchored in foundational questions that do not shift with each iteration: What human problem does this technology address? Whose interests does it advance, and whose does it undermine? What principles—equity, dignity, sovereignty—must remain non-negotiable even as tools evolve? These questions provide coherence to adaptation, ensuring flexibility does not drift into inconsistency.

It is equally important to acknowledge the inherent uncertainty of these domains. There will be errors—of anticipation and omission. Some technologies will accelerate far beyond forecasts; others will stall or converge in unexpected ways. The test of governance is not whether such surprises occur—they inevitably will—but whether institutions possess the humility and capability to recognize them early and adjust before harm compounds.

To institutionalize such responsiveness, policy processes must embed mechanisms for reflection and renewal. Sunset clauses that mandate reconsideration, regulatory sandboxes that enable controlled experimentation, and transparent feedback loops that surface unintended consequences are not procedural add-ons; they are essential safeguards in an environment defined by rapid change.

Crucially, this approach builds legitimacy. Citizens and markets are more likely to trust governance that is candid about uncertainty yet firm in its values—willing to revise based on evidence, yet unwavering in its objectives. In a world where technical consensus can shift within months, it is this combination of principled clarity and operational humility that distinguishes credible institutions from those left behind.

The ultimate challenge, then, is not merely to keep pace with technological change, but to build a policy architecture capable of learning alongside it—adapting without losing coherence, and doing so without sacrificing inclusivity or democratic legitimacy. Policy cannot remain a monument fixed in time; it must become a living system, whose ability to self-correct is its greatest strength.

By embracing this philosophy, governance becomes more than reactive. It becomes anticipatory—anchored in enduring principles, yet agile enough to navigate uncertainty. In this lies the possibility not only of regulating technology, but of shaping its trajectory in service of the common good.

From Regulation to Stewardship

Emerging technologies introduce a deeper challenge: not only speed, but radical uncertainty. AI systems evolve beyond their training data; gene editing may trigger ecological effects we cannot yet model; quantum computing could disrupt cryptographic systems with far-reaching geopolitical consequences. Policymakers are required to make decisions today whose full implications cannot be reliably predicted.

Traditional regulation assumes known risks and stable causal relationships. But in these domains, risk itself is uncertain, and second-order effects often emerge only after deployment. Waiting for certainty invites irrelevance; acting prematurely risks locking in flawed choices.

This demands a shift in mindset—from static control to active stewardship. Frameworks must be iterative, provisional and transparent about their assumptions. They must create space for safe experimentation—through mechanisms such as regulatory sandboxes—while continuously monitoring real-world outcomes and revising accordingly. This requires institutional humility to acknowledge what is unknown, and the courage to act regardless, guided by constitutional values and empirical evidence.

The traditional image of policy is one of rule-making and enforcement: a regulator standing guard, sanctioning deviations from established norms. Emerging technologies render this model insufficient. Their pace and unpredictability require policymakers to move beyond static regulation towards stewardship.

Stewardship is not a dilution of authority; it is a deeper form of responsibility. It involves shaping the trajectory of technological systems before they crystallize into structures that are difficult to redirect. It means engaging not only with consequences after deployment, but with intentions and incentives at the design stage.

This requires policy to intervene earlier in the innovation cycle. Rather than waiting for harm to manifest at scale, stewardship calls for sustained engagement with researchers, entrepreneurs and civil society while technologies are still evolving. The goal is not to control invention, but to ensure that decisions embedded in code and architecture reflect societal priorities.

In practice, this may involve regulatory sandboxes, adaptive licensing and experimental frameworks. These tools allow real-world testing under supervision, offering policymakers insights into likely consequences before large-scale rollout. They bridge the gap between legal abstraction and empirical reality.

Yet stewardship cannot rest on process alone. It must be anchored in values. Efficiency, growth and competitiveness remain important—but so must equity, transparency and human dignity. A steward asks not only whether a technology works, but whether it works for the society it claims to serve.

This perspective transforms the nature of policymaking. It shifts from adversarial oversight to shared responsibility. Regulators become co-authors of outcomes, rather than mere enforcers of compliance. It also demands humility: an acceptance that no institution, however capable, can anticipate every consequence.

Stewardship further implies continuity. Technologies evolve across electoral cycles and administrative tenures. Institutions must therefore develop memory—sustaining dialogue, learning and adaptation beyond immediate political horizons. In India, where administrative turnover is frequent, this presents a structural challenge requiring deliberate institutional innovation.

Above all, stewardship recognizes that governance is not an obstacle to progress, but part of its architecture. When designed well, it guides markets towards long-term resilience rather than short-term gain, and ensures that innovation strengthens rather than erodes public trust.

This is the deeper meaning of stewardship: to see policy not merely as a constraint, but as a strategic instrument in shaping national destiny.

Anticipating Convergence

Emerging technologies rarely remain confined to the domains in which they originate. Artificial intelligence, once the preserve of computer science, now shapes agriculture, healthcare and criminal justice. Blockchain spans finance, logistics and identity systems. Biotechnology increasingly intersects with data science and personalized medicine. Convergence, therefore, is not an exception; it is the rule.

Traditional governance structures struggle to keep pace with this reality. Policies are still drafted within vertical silos—reflected in constructs such as the 'India Stack'—with ministries, regulators and expert committees organized by sector. Each interprets technology through its own lens, producing frameworks that often overlap, conflict or leave gaps when technologies intersect in practice.

The result is fragmentation by design. A startup building AI-enabled diagnostics, for instance, must navigate health regulations, data protection norms and digital platform rules—each conceived independently of the others. This complexity privileges large incumbents with compliance capacity, while discouraging smaller innovators.

More critically, siloed policymaking obscures systemic consequences. A privacy rule designed for e-commerce may constrain medical research. A digital identity standard may inadvertently determine welfare eligibility. Without an integrated perspective, policy risks shifting burdens across sectors or creating unintended incentives.

Addressing convergence requires more than coordination; it demands institutional redesign. Inter-ministerial task forces, though common, are often episodic and procedural. What is needed is sustained capability—to map technological ecosystems, test policy coherence and engage across domains. This is not merely structural reform, but the cultivation of shared language and strategic foresight.

Foresight itself must become a core discipline. Tools such as horizon scanning, scenario planning and systems modelling allow policymakers to explore how technologies may recombine and what implications may follow. These are not predictive exercises, but structured engagements with uncertainty.

An integrated approach must also recognize that convergence amplifies both benefits and risks. An AI-enabled biobank may transform medicine, but also enable genetic discrimination. A universal digital payments system can drive inclusion, yet increase systemic exposure to cyber risk. Policy must therefore balance system-wide resilience with sectoral efficiency.

Importantly, anticipating convergence is not about constraining innovation. It is about ensuring that regulation remains fit for purpose as technologies evolve. It is, at its core, an act of strategic humility—recognizing that the most consequential effects often emerge at the intersections.

In a country of India's scale, this becomes imperative. A blind spot in one domain can propagate rapidly across others, affecting millions. To understand convergence is to look beyond institutional boundaries and engage with how technologies actually live within society.

In doing so, governance shifts from reactive patchwork to anticipatory design—a transformation as difficult as it is necessary in an age of relentless technological change.

The Risk of Over-Correction

Policy often operates under the pressure of urgency. Public sentiment, media cycles and geopolitical competition can compel swift regulatory action. While responsiveness is essential, haste introduces its own risks—chief among them, over-correction.

Over-correction is typically driven by fear: fear of falling behind, of public backlash, or of appearing indecisive. It manifests as sweeping bans, disproportionate compliance burdens or wholesale adoption of foreign regulatory models without local adaptation.

Such responses may offer immediate political reassurance, but they often impose long-term costs. Excessive regulation can deter domestic innovation, push startups offshore and entrench incumbents who alone can absorb compliance burdens. Paradoxically, it increases dependence rather than reducing it.

History offers familiar examples. Attempts to tightly regulate encryption have weakened domestic cybersecurity ecosystems. Overly rigid data localization proposals risk undermining cloud infrastructure critical to small businesses, without significantly enhancing security.

Recognizing this pattern requires separating signal from noise. The mere fact that another country has acted swiftly does not imply that India must replicate its approach. National contexts differ—in infrastructure, social norms and strategic priorities.

Deliberation, therefore, must not be mistaken for delay. Careful consultation, empirical analysis and scenario testing can reveal when urgency is justified and when restraint better serves long-term interests. Strategic patience is an underappreciated virtue in policymaking.

At the same time, policy must guard against capture. Powerful actors may invoke national security or public welfare to advance narrow interests. Transparency and evidence-based justification are essential counterweights.

The objective is not to avoid intervention, but to ensure that it is proportionate, context-specific and reversible. Over time, layers of over-correction create regulatory inertia—locking in inefficiencies and creating constituencies resistant to reform.

Restraint, therefore, is not passivity. It is discipline—rooted in confidence, not imitation. By resisting over-correction, India preserves space to shape its own innovation trajectory, its own narrative and its own strategic path.

Representation in Policymaking

Technologies that reshape society must be governed through processes that reflect that society. Yet policymaking often draws disproportionately from central ministries, industry bodies and urban technical experts, while voices from rural communities, small enterprises and marginalized groups remain peripheral.

This imbalance is not merely procedural—it shapes outcomes. Policies designed without engagement from the margins may fail to anticipate their impact on livelihoods, cultural contexts or local institutions. The result is governance that is technically sound but socially fragile.

True representation must be deliberate. Consultation must extend beyond organized stakeholders to include those whose interests are diffuse or historically excluded. Civil society, local governments and informal sector participants often surface insights invisible to centralized processes.

Moreover, representation is not static. As technologies change, so do the groups they affect. Periodic reassessment of stakeholder inclusion is therefore essential. Accessibility matters as well. Policies drafted solely in technical or legal language exclude meaningful participation. Regional translations, public consultations and digital engagement platforms can widen the circle.

Inclusion, however, introduces complexity. Broader participation slows decision-making and surfaces conflicting perspectives. Policymakers must therefore invest in processes that facilitate synthesis, mediate trade-offs and ensure transparent outcomes.

At the same time, diversity acts as a safeguard against capture. When policymaking is dominated by a narrow group, it risks aligning with their interests rather than the public good. Broader participation diffuses influence and strengthens legitimacy.

At its core, representation affirms that governance is collective. Technologies that affect every domain of life cannot be responsibly governed by a narrow elite. Expanding participation makes policy not only more legitimate, but also more resilient.

In a country as plural as India, this is not idealism—it is necessity. The future cannot be authored from a single vantage point, however expert. Only through structured, intentional inclusion can policymaking keep pace with technologies that affect every citizen differently.

Contours of a Blind Spot: India's Public Policy Value Chain

Policy, at its best, is a society's attempt to reason with its own complexity. It is the disciplined craft of translating conflicting interests and imperfect knowledge into collective direction. Yet in India—as in many large democracies—the lived practice of policymaking often feels brittle, hurried and partial. Grand designs are articulated on paper; consultation windows open and close; rules are drafted with care; and yet, for those beyond Delhi's corridors or state capitals, policy can appear remote, procedural or puzzlingly inconsistent.

This is not for want of dedication. Many within government and advisory circles work under genuine pressure, balancing political constraints, fiscal realities and social demands. The question is more structural: why does the value chain itself so often fail to match the scale and dynamism of the society it seeks to guide—from problem identification to drafting, consultation, enactment, implementation, monitoring and revision?

Part of the answer lies in the past—in institutions and habits formed when policy moved more slowly than markets, and when knowledge was concentrated in predictable centres. But it also lies in the present, shaped by the speed and opacity of emerging technologies, whose implications are neither fully understood by their creators nor easily anticipated by those tasked with governing them.

To critique the Indian policy value chain, therefore, is not to argue for more control or less ambition. It is to argue for humility in the face of uncertainty, discipline in the face of complexity, and an institutional design that learns rather than merely reacts. In this century, the cost of getting policy wrong is rising—but so is the cost of standing still.

From Problem to Policy: The Narrow Gate

Every policy begins with the recognition of a problem that warrants collective response. Yet in practice, only certain problems pass through the narrow gate into formal agendas. Judicial orders, media investigations and lobbying by organized sectors often determine what becomes urgent. Structural challenges that evolve gradually—the absence of safety nets for informal workers, regional disparities in digital access or the environmental consequences of industrial growth—rarely command the same immediacy.

This gatekeeping is neither malicious nor conspiratorial; it is structural. It privileges what can be distilled into headlines or executive summaries, what aligns with prevailing political narratives or fiscal cycles. Problems without a clear 'owner' or those affecting constituencies with limited organizational voice tend to remain at the threshold.

In a country as diverse as India, this upstream narrowing has real consequences. It systematically over-represents urban, formal, English-speaking interests, while under-representing those at the intersection of informal labour, caste, language and geography. It helps explain why policy often seems to echo certain voices more than others, and why the perception of distance between policymaking and everyday life persists.

The first discipline of a reflective policy value chain, therefore, is to widen this gate. Not by flattening all priorities, but by investing in deliberate problem discovery: local surveys, frontline administrative data, partnerships with regional universities and structured mechanisms to identify issues that affect many but are voiced by few.

Policy Design: The Tyranny of Templates

Once a problem crosses the threshold, the instinct is often to reach for familiar instruments: licences, quotas, caps, mandatory disclosures, compliance portals. These are not arbitrary choices; they are rooted in long institutional experience. But they carry an implicit model of society as a set of linear cause-and-effect relationships, where behaviour can be predictably reshaped through prescriptive rules.

Emerging technologies challenge this model at its corc. Artificial intelligence, genome editing, quantum computing, blockchain systems and autonomous technologies do not behave like static products; they evolve, adapt and interact in ways even their creators cannot fully anticipate. A rule drafted in 2026 may be outpaced by technical change before it is fully implemented.

Yet policy design often remains anchored in the belief that completeness produces control. Drafters attempt to anticipate every possible misuse or loophole, resulting in frameworks whose complexity becomes a barrier (e.g., the inherent complexity of the GST regime with a multitude of rates and slabs that is now being re-organized, but not fully).Complexity, however, has uneven effects. Large firms can absorb it by hiring expertise; smaller actors and ordinary citizens often encounter it as a point of friction—where compliance becomes avoidance.

A more resilient approach would invert the question. Not 'how do we exhaustively regulate what we know?' but 'how do we design frameworks that remain valid when confronted with what we do not yet know?' This requires anchoring policy in first principles—accountability, proportionality, equity—while embedding mechanisms for revision as evidence evolves.

Consultation: Between Form and Substance

Consultation is the moment when policy pauses to listen. India has made progress in formalizing this process: draft texts are published, submissions invited, and public engagement occasionally attempted. Yet in practice, consultation often remains narrow.

Written submissions, typically in English, privilege those with the capacity to engage in technical discourse. Industry bodies, large NGOs and a small set of think tanks dominate participation, while small entrepreneurs, community organizations and linguistic minorities remain underrepresented.

Even among participants, scepticism persists: what impact does consultation actually have? Ministries seldom provide detailed accounts of how feedback shaped final decisions. The process can feel procedural—a ritual of inclusion rather than an exercise in shared reasoning.

Emerging technologies magnify these gaps. Their abstract and technical nature makes meaningful participation difficult even for most well-educated citizens, concentrating influence among experts and corporate actors who speak the language of code, data and risk—but may not fully grasp how these systems shape everyday life.

A more substantive approach to consultation would go beyond soliciting written feedback. It would invest in explanatory materials in regional languages, convene deliberative forums that combine expert knowledge with lived experience and design processes that surface disagreement rather than prematurely smoothing it over. It would treat dissent not as obstruction, but as a signal of underlying complexity.

Drafting and Enactment: Complexity Compounded

Drafting transforms policy intent into legal architecture. India's tradition of detailed drafting is rooted in a legitimate desire for clarity and enforceability. Yet in the context of complex, adaptive systems, detail can paradoxically produce opacity. Lengthy provisions, nested exceptions and cross-references to multiple statutes create texts that few non-lawyers—and often even practitioners—can easily interpret.

This complexity is not accidental. Political negotiation tends to add rather than subtract. Parliamentary committees, stakeholder interventions and legal risk mitigation often layer new clauses instead of revisiting core structures. The result is legislation that can be internally inconsistent, open to judicial reinterpretation and unevenly applied across contexts.

Enactment, moreover, rarely marks the end of design. Much of a policy's operational reality is shaped later through rules, notifications and guidelines. These are often drafted under compressed timelines, with limited external scrutiny. Stakeholders who engaged meaningfully during the legislative phase frequently discover that implementing rules alter incentives in ways the primary law did not anticipate or signal.

A more reflective drafting process would privilege simplicity anchored in principle, ensure clear delegation of rule-making authority, and embed structured revision cycles. It would ask not only whether the text covers every conceivable scenario, but whether those tasked with implementation—and those subject to compliance—can reasonably interpret and act upon it.

Implementation: Between Ambition and Capacity

Implementation is where policy ideals encounter reality—administrative bandwidth, local political economies, technological infrastructure and uneven state capacity. Even well-designed policies can falter when frontline agencies lack trained personnel, IT systems fail to interoperate, or local officials operate under conflicting incentives.

Emerging technologies amplify these challenges. Rules that appear neutral on paper may become exclusionary in practice if small enterprises cannot afford compliance systems, or if local governments lack the digital infrastructure to operationalize them. Data-driven regulation may produce dashboards rich in numbers, but thin in context.

Coordination across central, state and municipal levels remains a structural constraint. Policies assume vertical alignment, yet fiscal asymmetries, political variation and institutional fragmentation produce divergent interpretations and uneven enforcement. These differences are rarely visible in headline metrics but become evident in who benefits—and who is silently excluded.

Implementation must therefore be integral to design, not an afterthought. It requires realistic timelines, sustained investment in institutional capacity, and feedback mechanisms that allow for mid-course correction before minor gaps expand into systemic failures.

Monitoring and Revision: The Weakest Link

No policy, however carefully designed, can anticipate all outcomes. The ability to evaluate impact, identify unintended consequences and revise accordingly is what transforms policy from a static decree into a living system. Yet this remains the weakest link in India's policy value chain.

Ex-post evaluations are sporadic. Data collection is fragmented. Sunset clauses that mandate periodic review are rare. Once policies are enacted, institutional incentives shift from questioning to defending them.

The consequences are, thus, serious in fast-evolving fields: rules drafted for initial conditions can entrench market structures, deter new entrants or embed biases into systems that scale rapidly. Without structured revision, policies risk drifting further from both technological realities and social needs.

Normalizing revision is therefore essential. It must be seen not as a sign of failure, but as evidence of responsibility. This requires independent evaluation mechanisms, transparent publication of findings, and explicit pathways linking evidence to redesign.

The Expertise Gap: Technology and Policy

A final blind spot compounds these challenges: the limited presence of deep technological expertise within policymaking institutions. Laws governing AI, blockchain or synthetic biology are often shaped in environments where few participants can rigorously interrogate technical claims, understand system architectures or anticipate feedback loops.

This gap shifts agenda-setting power outward—to large incumbents, external consultants and specialized networks whose incentives may not fully align with public purpose. It also leaves policymakers vulnerable to cycles of hype, fear or geopolitical pressure, rather than grounded evidence.

Bridging this gap requires institutional redesign. Dedicated policy labs within ministries, cross-disciplinary fellowships, secondments of technologists into government, and career pathways that reward hybrid expertise are no longer optional—they are foundational. Without this capability, policymaking risks becoming reactive: imitative rather than original, and speculative rather than strategic.

To critique the Indian policy value chain is not to deny its necessity or to idealize the possibility of perfect foresight. It is to recognize that in an age defined by technologies we do not fully understand, and systems whose consequences we cannot fully predict, the cost of brittle, one-time policymaking grows steadily higher.

A sovereign state is more than a regulator; it is a trustee of its citizens' freedoms. In the digital age, this responsibility extends into invisible domains—where data is harvested, decisions are automated, and power is exercised without visibility. To govern without recognizing this is to confuse administration with stewardship.

Governments can build platforms, mandate compliance and issue regulations. But unless these instruments are directed towards safeguarding dignity and protecting agency, they remain tools without moral purpose.

When policy is built on myths rather than math, the consequence is not merely inefficiency—it is the erosion of sovereignty itself. Digital power is inherently asymmetric: those who define standards, write protocols or control infrastructure shape outcomes, regardless of where formal authority resides.

If policy mistakes visibility for control, sovereignty becomes a ritual, not a reality. The state can pass rules, issue directives and mandate compliance, but if the foundations are outside our reach, the rules govern only the surface.

India has shown this pattern repeatedly. Our discourse often conflates symbolic control with structural control. We celebrate domestic digital champions without asking whether the underlying infrastructure—cloud layers, cryptographic standards, semiconductor supply chains—is within our command. We draft regulations that demand access without building enforcement capacity or resilience.

In doing so, we regulate form while ceding substance. The intention may be patriotic, but the execution often leaves sovereignty hollowed out.

Nowhere is this tension clearer than in the debate on data localization. The issue is often framed in geographic terms: whether data resides within India or beyond its borders. Geography matters, but the deeper question is who controls the locks—the encryption keys, the protocols and the algorithms that store, process and move that data. Local storage without control over the underlying stack is akin to building a vault while outsourcing the locksmith.

Estonia, a country with a fraction of India's population, understood this distinction two decades ago. Through its X-Road digital backbone, it ensured that sovereignty was not defined merely by the location of data, but by the ability of citizens and the state to securely control, exchange and authenticate it without dependence on foreign platforms. It was not scale but clarity of design that made sovereignty meaningful.

The challenge in India is compounded by a political appetite for narrative. We have become adept at projecting world-stage achievements—the 'largest', the 'fastest', the 'first'. UPI is celebrated globally as the crown jewel of digital public infrastructure; Aadhaar is hailed as the largest identity project in the world. These narratives make for powerful PR, but they obscure lived reality. UPI's success coexists with fraud and limited recourse. Aadhaar's scale coexists with exclusion errors that can deny essential services to the most vulnerable. Sovereignty is not realized through scale alone; it is realized when scale is matched with safeguards that protect dignity, especially when systems fail.

Policymakers are not unaware of these risks. The deeper issue is that myths are easier to govern than math. A myth offers the illusion of immediate sovereignty—a rule announced, a directive issued, a compliance certificate obtained. Math demands something more demanding: the slow construction of building capabilities, creating standards and designing redundancies. It forces the recognition that sovereignty cannot be declared; it must be engineered.

India has often chosen the former—not out of neglect, but because the political economy rewards short-term optics over long-term architecture.

The costs, however, are not abstract. They are borne directly by citizens. A citizen may believe they own their bank account, data and digital identity. In reality, these may be contingent rights—one that survives only as long as foreign platforms or global sanction regimes allow it. Sovereignty without substance erodes freedom at its roots, leading to a nation where rights are fragile, suspended at will and dependent on invisible systems beyond our control.

Other jurisdictions illustrate a different path. The European Union, for instance, used GDPR not only as a data protection framework but also as a lever of sovereignty. By setting stringent global privacy standards and enforcing them with billion-euro fines, against even the largest corporations, the EU demonstrated that a well-crafted policy rooted in math—legal clarity, enforceability and economic leverage—can rebalance power even in the absence of technological dominance. Regulation, in this case, was not merely compliance but strategy.

India has yet to fully deploy regulation in this manner.

The tragedy is that the gap is not one of imagination or talent, but of sustained ambition. Myth-driven policymaking is politically easier as it offers immediate reassurance; math-driven policymaking is strategically necessary as it builds long-term strength. Each time the former is chosen, the latter is deferred—and in digital systems, delay compounds disadvantage. Falling behind in standards, cryptography or hardware ecosystems makes future catch-up exponentially harder. The comfort of myth today becomes the cost of dependency tomorrow.

Correcting this trajectory requires more than incremental reform. It demands a cultural shift in how policy is conceived. Regulators must be willing to ask hard questions about dependencies, identify hidden chokepoints and stress-test sovereignty against real scenarios—sanctions, cyberattacks, infrastructure withdrawal—rather than assuming compliance equates to resilience. It also requires embedding technologists, mathematicians and systems architects within the policymaking process, ensuring that laws are grounded in the architecture of reality rather than the comfort of analogy.

Above all, it requires ambition. Digital sovereignty cannot be a defensive posture; it must be a proactive strategy. We cannot settle for being custodians of appearances. India must design standards, create intellectual property and build infrastructure that others depend upon. Estonia achieved this through design clarity; the EU through regulatory will. India possesses both scale and talent—but unless myths yield to math, sovereignty will remain declaratory and freedom conditional.

A resilient policy architecture begins with humility: widening the gate through which problems are recognized, designing frameworks for adaptability, broadening consultation beyond familiar voices, drafting for clarity rather than exhaustiveness, implementing with attention to local capacity and revising not reluctantly but routinely.

Above all, it requires investing in institutional expertise equal to the complexity of the world it seeks to govern. For policy remains, at its heart, an act of collective imagination—deciding not only what must be done, but why, and remaining open to learning, listening and course correction.

Taken together, these principles—recognizing hidden architectures of power, shifting from regulation to stewardship, anticipating convergence, governing under uncertainty, resisting over-correction and ensuring representation—outline a vision of governance suited to the age of emerging technologies. Not reactive, but anticipatory; not rigid, but adaptive; not closed, but inclusive.

In embracing them, policy becomes more than a response to disruption. It becomes an instrument through which a nation consciously shapes its technological future. That is the task before us: to build governance that learns as technologies learn, guided not by fear or imitation, but by confidence in our values and purpose.

To reclaim it requires courage: the willingness to slow down in an age that rewards speed, to question consensus when imitation is easier and to recognize that true sovereignty lies not in keeping pace with others, but in deciding, from first principles, what must be built, protected or restrained.

If there is one lesson the digital age demands of India, it is this: sovereignty cannot be borrowed, and freedom cannot be outsourced. We will either build systems grounded in our own ambition, or inhabit systems designed elsewhere. The choice is not merely between myth and math—it is between dependency and dignity.

Digital sovereignty is not a slogan but a moral obligation to ensure that citizens of this country enjoy the same certainty of freedom online that earlier generations fought for in the political realm.

8

Freedom Is a System We Must Build

We hope what remains from this book is not merely a catalogue of risks or technological anxieties, but a recognition of the kind of century now unfolding before us and the seriousness with which India must respond to it. The argument running through these chapters emerges not from pessimism, but from the belief that a civilization of India's scale, complexity and historical depth cannot afford to drift passively through the defining technological transition of the modern era while allowing others to shape the architecture within which its future will operate.

The world that shaped the late 20th century has become more fragmented, contested and unstable. The assumptions that underpinned globalization, institutional trust and technological openness are weakening under the pressure of geopolitical rivalry, supply-chain disruptions, economic insecurity and concentrated technological power. Influence increasingly flows not merely through territory, military capability or capital accumulation. It now flows through operating systems, cloud infrastructure, semiconductor ecosystems, artificial intelligence, standards-setting frameworks and the invisible protocols through which modern societies function every day.

In such a world, dependence acquires strategic consequences. Nations that fail to build meaningful control over critical layers of intelligence, infrastructure and coordination eventually find their economic choices, diplomatic flexibility and social resilience constrained by systems designed elsewhere. This concern is already visible in India's overwhelming dependence on external operating systems, cloud infrastructure, imported hardware architectures, proprietary software ecosystems and digital platforms that increasingly mediate commerce, communication, advertising and public discourse itself.

India has undoubtedly achieved remarkable success in building population-scale digital infrastructure and enabling adoption at extraordinary speed. Yet adoption alone does not translate into strategic influence or digital self-sufficiency. The deeper challenge before India is institutional. It concerns whether the country can build durable state capacity, resilient supply chains, advanced research ecosystems and governance frameworks capable of operating at the velocity and complexity of the digital century.

Indian intellectual traditions, at their strongest, never treated introspection as weakness. The Upanishadic instinct of refinement through relentless examination, Kautilya's insistence on strategic foresight and long-horizon statecraft, and the broader Indic understanding that wisdom emerges through contestation rather than conformity all point towards a culture that once valued scrutiny as an instrument of endurance.

That matters profoundly today because the systems now being built will shape not merely markets or industries, but the conditions under which freedom itself will be experienced. Modern societies increasingly rely on digital infrastructures to mediate everyday existence. Banking systems determine access to savings and wages. Identity architectures shape access to welfare, mobility and citizenship. Health systems, logistics networks, educational records and communication platforms have become deeply interdependent layers of social organization.

As these systems become more integrated and algorithmically governed, their failures cease to remain narrowly technical. A systems outage, cybersecurity breach or failed authentication layer can rapidly cascade into disruptions affecting livelihoods, healthcare, food distribution or public order. For this reason, governance in the digital century carries obligations that extend beyond innovation or efficiency.

Citizens experience the state most directly through the reliability of the systems they depend upon every day. A pension credited on time, a payment processed without friction, a hospital network functioning during a crisis, an identity system recognizing a citizen without humiliation or exclusion—these moments collectively constitute the lived legitimacy of institutions. The durability of a digital republic rests on whether ordinary life can proceed without fear that unseen systems may suddenly become sources of instability.

Freedom in the digital century therefore demands greater institutional seriousness. It cannot survive merely as a constitutional abstraction while economic life, communication systems and informational visibility become increasingly centralized within opaque technological structures. A society, where citizens become fully transparent to institutions while institutional power itself remains shielded from scrutiny, gradually normalizes imbalance as governance.

The preservation of freedom depends not only on legal rights, but on the design principles embedded into digital systems themselves: accountability, interoperability, privacy, contestability and the distribution of power.

India's challenge over the coming decades will lie in aligning technological ambition with institutional maturity. Advanced technologies alone will not secure sovereignty unless accompanied by sustained investment in scientific capability, research depth, industrial capacity, cybersecurity resilience, educational reform and administrative competence. Nations that aspire to shape the digital order must cultivate the ability to build foundational technologies, govern complex systems responsibly and sustain public trust through reliability rather than rhetoric.

Such capabilities rarely emerge through short-term enthusiasm or symbolic declarations. They require institutions capable of learning, adapting and investing across generational timescales. They require universities that produce original inquiry rather than credential inflation, cities capable of governing complexity, regulators that evolve alongside technological risk and political systems willing to prioritize strategic resilience over immediate applause.

A quarter of this century has already passed. The infrastructures, standards and behavioural defaults established during the next two decades will likely endure far beyond them. Once dependence hardens into architecture, reversing it becomes extraordinarily difficult.

The urgency of this moment therefore lies not in fear, but in recognition: recognition that sovereignty, privacy and identity are no longer peripheral concerns in the digital age, but foundational conditions upon which political autonomy, economic stability and civic freedom increasingly depend.

What remains within collective control, however, is the seriousness with which societies prepare for it. The responsibility before India is therefore not merely technological or economic, but civilizational: to ensure that freedom, sovereignty and institutional trust survive the systems through which the future will increasingly be governed.

Endnotes

1 Eichengreen, Barry, *Exorbitant Privilege: The Rise and Fall of the Dollar and the Future of the International Monetary System*, Oxford University Press, Oxford, 2011.

2 Haidar, Suhasini, 'Alaskan Winds, India and the Trump-Putin Summit', *The Hindu*, 19 August 2025, https://www.thehindu.com/opinion/lead/alaskan-winds-india-and-the-trump-putin-summit/article69948313.ece. Accessed on 21 May 2026.

3 Mukherjee, Andy, 'India's 50% Tariff Is a US Sanction in Disguise', *Bloomberg*, 8 August 2025, https://www.bloomberg.com/opinion/articles/2025-08-07/india-s-50-tariff-is-a-us-sanction-in-disguise. Accessed on 21 May 2026.

4 'China Reportedly Chokes Fertiliser Supply to India Amid Trade Tensions, Puts Cropping Season at Risk', *Moneycontrol*, 26 June 2025, https://www.moneycontrol.com/news/business/economy/china-reportedly-chokes-fertiliser-supply-to-india-amid-trade-tensions-puts-cropping-season-at-risk-13180509.html. Accessed on 21 May 2026.

5 Doval, Pankaj, and Sidhartha, 'China Targets India's Manufacturing Sector, Delays Machinery Delivery, Pulls iPhone Engineers', *The Times of India*, 3 July 2025, https://timesofindia.indiatimes.com/business/india-business/china-targets-indias-mfg-sector-delays-machinery-delivery-pulls-iphone-engrs/articleshow/122214492.cms. Accessed on 21 May 2026.

6 McCarthy, Simone, and Aishwarya S. Iyer, 'Herders on Front Line of India's Himalayan Dispute with China Say They're Losing Grazing Land – and a Way of Life', *CNN*, 16 April 2024, https://edition.cnn.com/2024/04/16/china/ladakh-india-china-border-intl-hnk. Accessed on 21 May 2026.

7 'Paris Economic Declaration', 16 July 1989, para. 52, 'Drug Issues', https://g7.utoronto.ca/summit/1989paris/communique/index.html. Accessed on 21 May 2026.

8 '12 U.S.C. 1829b, Retention of Records by Insured Depository Institutions', https://www.govinfo.gov/content/pkg/USCODE-2020-title12/pdf/USCODE-2020-title12-chap16-sec1829b.pdf. Accessed on 21 May 2026.

9 'Interview: Mike McDonald', *Frontline*, PBS, https://www.pbs.org/wgbh/pages/frontline/shows/drugs/interviews/mcdonald.html. Accessed on 21 May 2026.

10 Ibid.

11 Sanger, David E., 'The World; Money Laundering, New and Improved', *The New York Times*, 24 December 1995, https://www.nytimes.com/1995/12/24/weekinreview/the-world-money-laundering-new-and-improved.html. Accessed on 21 May 2026.

12 Zill, Oriana, and Lowell Bergman, 'U.S. Business & Money Laundering', *Frontline*, PBS, https://www.pbs.org/wgbh/pages/frontline/shows/drugs/special/us.html. Accessed on 21 May 2026.
13 'Interview: Greg Passic', *Frontline*, PBS, https://www.pbs.org/wgbh/pages/frontline/shows/drugs/interviews/passic.html. Accessed on 21 May 2026.
14 Barnet, Todd, 'Legal Fiction and Forfeiture: An Historical Analysis of the Civil Asset Forfeiture Reform Act', *Duquesne Law Review*, Vol. 40, 2001, 77, https://dsc.duq.edu/dlr/vol40/iss1/4. Accessed on 21 May 2026.
15 Reserve Bank of India, *Master Circular on Maintenance of Deposit Accounts, UCBs*, RBI/2009-10/78UBD.BPD(PCB) MC.No 13/13.01.000/2009-10, July 2009, Section 2.1.1, https://www.rbi.org.in/commonperson/English/Scripts/Notification.aspx?Id=598. Accessed on 21 May 2026.
16 Ibid, Section 2.2, https://www.rbi.org.in/commonperson/English/Scripts/Notification.aspx?Id=598. Accessed on 21 May 2026.
17 Financial Action Task Force on Money Laundering, *Annual Report 1989–1990*, 2 July 1990, pp. 31–32, https://www.fatf-gafi.org/content/dam/fatf-gafi/annual-reports/1990%20ENG.pdf. Accessed on 21 May 2026.
18 Government of India, *Prevention of Money Laundering Act, 2002*, https://web.archive.org/web/20171118111906/https://dea.gov.in/sites/default/files/moneylaunderingact.pdf. Accessed on 25 May 2026.
19 Financial Intelligence Unit – India, Ministry of Finance, Government of India, 'PML (Maintenance of Records) Rules 2005', https://fiuindia.gov.in/files/AML_Legislation/notification.html. Accessed on 25 May 2026.
20 Ibid, Rule 9.
21 Ibid, Rule 9A.
22 Central Registry of Securitisation Asset Reconstruction and Security Interest of India (CERSAI), *Central KYC Records Registry Operating Guidelines*, Version 1.4.1, p. 10, https://www.ckycindia.in/ckyc/assets/doc/Operating_Guidelines_version1.4.1.pdf. Accessed on 25 May 2026.
23 Department of Telecommunications, Government of India, 'List of Acceptable Documents as Proof of Identity and Proof of Address from General Public in the Country except in J&K, North East and Assam Service Areas', https://web.archive.org/web/20170110183111/https://dot.gov.in/sites/default/files/2016_11_18%20POIA-AS-II.pdf. Accessed on 25 May 2026.
24 'UIDAI Asks CSC to Close Aadhaar Services Business', *Moneylife*, 8 February 2018, https://www.moneylife.in/article/enormous-number-of-complaints-of-corruption-and-enrolment-process-violations-against-aadhaar-enrolment-and-update-centres-under-csc-egov-says-uidai-while-asking-its-sibling-to-close-aadhaar-services-business/53017.html. Accessed on 21 May 2026.
25 Ramanathan, Ajay, 'High Costs, Weak Tech Infra Makes KYC a Challenge: Fintechs', *The Financial Express*, 7 February 2024, https://www.financialexpress.com/business/banking-finance-high-costs-weak-tech-infra-makes-kyc-a-challenge-fintechs-3386319/. Accessed on 21 May 2026.
26 Government of India, *The Prevention of Money-Laundering Act, 2002, Enforcement*

Directorate,, sec. 73, p. 38, https://web.archive.org/web/20220518221143/https://enforcementdirectorate.gov.in/sites/default/files/Act&rules/THE%25252520PREVENTION%25252520OF%25252520MONEY%25252520LAUNDERING%25252520ACT,%252525202002.pdf. Accessed on 25 May 2026.

27 Department of Banking Supervision, Reserve Bank of India, https://www.rbi.org.in/commonman/English/scripts/deptofbs.aspx. Accessed on 21 May 2026.

28 Reserve Bank of India, *KYC Master Direction*, Part IV, sec. 37, https://www.rbi.org.in/commonman/English/scripts/notification.aspx?id=2607. Accessed on 21 May 2026.

29 Rai, Suyash, 'Economic Development and Digital Transformation: Learning from the Experience of Aadhaar and Financial Inclusion in India', *XKDR Forum Working Papers*, No. 35, xKDR, 2024, pp. 17–19.

30 Ibid, p. 30.

31 US Department of State, 'Ukraine and Russia Sanctions', https://www.state.gov/division-for-counter-threat-finance-and-sanctions/ukraine-and-russia-sanctions. Accessed on 21 May 2026.

32 *EU Sanctions Map*, https://www.sanctionsmap.eu/#/main. Accessed on 21 May 2026.

33 *EU Sanctions Tracker*, https://data.europa.eu/apps/eusanctionstracker. Accessed on 21 May 2026.

34 Sukumar, Tanay, 'How Indians Move within the Country in Charts', *Mint*, 2 December 2023, https://www.livemint.com/economy/in-charts-how-indians-move-within-the-country-11678642949279.html. Accessed on 21 May 2026.

35 Narayanan, Arvind, and Sayash Kapoor, *AI Snake Oil: What Artificial Intelligence Can Do, What It Can't, and How to Tell the Difference*, Princeton University Press, Princeton, 2024, p. 1.

36 Ibid. Out of the several examples cited by Arvind Narayanan and Sayash Kapoor, behaviour prediction based on video clips during hiring, criminal risk prediction tools are egregious examples of AI snake oil.

37 Mantegna, R.N., 'Hierarchical Structure in Financial Markets', *The European Physical Journal B*, Vol. 11, No. 1, 1999, pp. 193–97.

38 Cardoso, Mário, Pedro Saleiro, and Pedro Bizarro, 'LaundroGraph: Self-Supervised Graph Representation Learning for Anti-Money Laundering', *Proceedings of the 3rd ACM International Conference on AI in Finance (ICAIF '22)*, ACM, New York, 2022, https://doi.org/10.1145/3533271.3561727. Accessed on 21 May 2026.

39 Li, Xiang, Wen Zhang, Jiuzhou Xi, and Hao Zhu, 'HGsuspector: Scalable Collective Fraud Detection in Heterogeneous Graphs', in Proceedings of ACM Conference, ACM, New York, 2018, Article 4, https://doi.org/10.475/123_4. Accessed on 21 May 2026.

40 Kanezashi, Hiroki, Toyotaro Suzumura, Xin Liu, and Takahiro Hirofuchi, 'Ethereum Fraud Detection with Heterogeneous Graph Neural Networks', *Proceedings of ACM Conference*, ACM, New York, 2022, https://arxiv.org/abs/2203.12363. Accessed on 21 May 2026.

41 IFTAS, 'Cloud Services', https://www.iftas.in/Cloud_Services. Accessed on 21 May 2026.

42 Hellmann, Aline Gazola, 'How Does Bolsa Familia Work?', *Inter-American Development Bank*, September 2015, https://publications.iadb.org/en/publications/english/viewer/How-Does-Bolsa-Familia-Work-Best-Practices-in-the-Implementation-of-Conditional-Cash-Transfer-Programs-in-Latin-America-and-the-Caribbean.pdf. Accessed on 21 May 2026.

43 'On the Industrial History of the United Kingdom', *European Route of Industrial Heritage (ERIH)*, https://web.archive.org/web/20210731175913/https://www.erih.net/how-it-started/industrial-history-of-european-countries/united-kingdom. Accessed on 21 May 2026.

44 New Jersey Historical Records Survey, *Project Copy of the Calendar of the S.U.M. Collection of Manuscripts*, http://patersongreatfalls.org/letters/Intro_NJ.pdf. Accessed on 21 May 2026.

45 Act of 3 April 1948, *European Recovery Act (Marshall Plan)*, Enrolled Acts and Resolutions of Congress, 1789–1996, General Records of the United States Government, Record Group 11, National Archives, https://www.archives.gov/milestone-documents/marshall-plan. Accessed on 21 May 2026.

46 Kennedy, Scott, 'Made in China 2025', *Critical Questions*, Center for Strategic and International Studies (CSIS), 20 November 2025, https://www.csis.org/analysis/made-china-2025. Accessed on 21 May 2026.

47 International Energy Agency (IEA), *Solar PV Global Supply Chains*, IEA, Paris, 2022, https://www.iea.org/reports/solar-pv-global-supply-chains. Accessed on 21 May 2026.

48 U.S. Geological Survey, *Data Release for Mineral Commodity Summaries 2022*, U.S. Geological Survey, 2022, https://doi.org/10.5066/P9KKMCP4. Accessed on 21 May 2026.

49 Bhandari, Konark, 'Is India "Ready" for Semiconductor Manufacturing?', *Carnegie India*, 23 May 2023, https://carnegieendowment.org/russia-eurasia/research/2023/05/is-india-ready-for-semiconductor-manufacturing. Accessed on 25 May 2026.

50 'The All New BHIM Baroda Pay', *Bank of Baroda*, https://www.bankofbaroda.in/writereaddata/images/pdf/new-upi-existing-users-final-1.pdf. Accessed on 25 May 2026.

51 IHS Electronics & Media, 'Apple iPhone 3G 8GB Teardown', *Electronics360*, 24 July 2008, https://electronics360.globalspec.com/article/3686/apple-iphone-3g-8gb-teardown. Accessed on 21 May 2026.

52 'The Cost of Compute: A $7 Trillion Race to Scale Data Centers', *McKinsey Quarterly*, 28 April 2025, https://www.mckinsey.com/industries/technology-media-and-telecommunications/our-insights/the-cost-of-compute-a-7-trillion-dollar-race-to-scale-data-centers. Accessed on 21 May 2026.

53 Pilz, Konstantin Friedemann, James Sanders, Robi Rahman, and Lennart Heim, 'Trends in AI Supercomputers', *ICML Workshop on Technical AI Governance (TAIG)*, 2025, https://openreview.net/forum?id=PEsWiS8jMT. Accessed on 21 May 2026.

54 Smith, Tim, 'Meta and DeepMind Alumni Raise €105m Seed Round to Build

OpenAI Rival Mistral', *Sifted*, 13 June 2023, https://sifted.eu/articles/mistral-openai-rival-105m-news. Accessed on 21 May 2026.
55 Heim, Anna, 'What is Mistral AI? Everything to know about the OpenAI competitor', *TechCrunch*, 9 September 2025, https://techcrunch.com/2025/07/18/what-is-mistral-ai-everything-to-know-about-the-openai-competitor/. Accessed on 21 May 2026.
56 Jiang, Albert Q., Alexandre Sablayrolles, Arthur Mensch, Chris Bamford, Devendra Singh Chaplot, Diego de las Casas, Florian Bressand, et al., 'Mistral 7B', 10 October 2023, https://arxiv.org/pdf/2310.06825. Accessed on 21 May 2026.
57 Gil, Elad, 'Discussion w Arthur Mensch, CEO of Mistral AI', *Elad Blog*, 22 March 2024, https://blog.eladgil.com/p/discussion-w-arthur-mensch-ceo-of. Accessed on 21 May 2026.
58 Shilov, Anton, 'DeepSeek's AI Breakthrough Bypasses Industry-Standard CUDA for Some Functions, Uses Nvidia's Assembly-Like PTX Programming Instead', *Tom's Hardware*, 28 January 2025, https://www.tomshardware.com/tech-industry/artificial-intelligence/deepseeks-ai-breakthrough-bypasses-industry-standard-cuda-uses-assembly-like-ptx-programming-instead. Accessed on 21 May 2026.
59 'Ascend PyTorch Codebase', *GitHub*, https://github.com/Ascend/pytorch. Accessed on 21 May 2026.
60 Ottinger, Lily, and Mary Clare McMahon, 'Can Huawei Take On Nvidia's CUDA? The True US v China AI Battle', *ChinaTalk*, 5 May 2025, https://www.chinatalk.media/p/can-huawei-compete-with-cuda. Accessed on 21 May 2026.
61 Mourales, Jowi, 'Huawei Introduces the Ascend 920 AI Chip to Fill the Void Left by Nvidia's H20', *Tom's Hardware*, 19 April 2025, https://www.tomshardware.com/pc-components/gpus/huawei-introduces-the-ascend-920-ai-chip-to-fill-the-void-left-by-nvidias-h20. Accessed on 21 May 2026.
62 Ottinger, Lily, and Mary Clare McMahon, 'Can Huawei Take On Nvidia's CUDA? The True US v China AI Battle', *ChinaTalk*, 5 May 2025, https://www.chinatalk.media/p/can-huawei-compete-with-cuda. Accessed on 21 May 2026.
63 'DeepSeek FAQ', *Stratechery*, 27 January 2025, https://stratechery.com/2025/deepseek-faq. Accessed on 21 May 2026.
64 Mhaske, Arnav, Harshit Kedia, Sumanth Doddapaneni, Mitesh M. Khapra, Pratyush Kumar, Rudra Murthy V, and Anoop Kunchukuttan, 'Naamapadam: A Large-Scale Named Entity Annotated Data for Indic Languages', 2023, https://arxiv.org/abs/2212.10168. Accessed on 21 May 2026.
65 Chadha, Harveen Singh, Anirudh Gupta, Priyanshi Shah, Neeraj Chhimwal, Ankur Dhuriya, Rishabh Gaur, and Vivek Raghavan, 'Vakyansh: ASR Toolkit for Low Resource Indic Languages', 2022, https://arxiv.org/abs/2203.16512. Accessed on 25 May 2026.
66 Arora, Gaurav, 'iNLTK: Natural Language Toolkit for Indic Languages', *Proceedings of the Second Workshop for NLP Open Source Software (NLP-OSS)*, Association for Computational Linguistics, 2020, pp. 66–71, http://dx.doi.org/10.18653/v1/2020.nlposs-1.10. Accessed on 25 May 2026.
67 Serrano-Cinca, C., B. Gutiérrez-Nieto, and L. López-Palacios, 'Determinants of

Default in P2P Lending', *PLoS ONE*, Vol. 10, No. 10, 2015, e0139427, https://doi.org/10.1371/journal.pone.0139427. Accessed on 25 May

68 Huang, Yangguang, 'Why Did the Peer-to-Peer Lending Market Fail in China?', *HKUST IEMS Thought Leadership Brief*, No. 61, Spring 2022, https://iems.hkust.edu.hk/assets/publications/thought-leadership-briefs/2022/hkustiems-huang-why-did-the-peer-to-peer-lending-market-fail-in-china-tlb61.pdf. Accessed on 25 May 2026.

69 'Guiding Opinions on Promoting the Sound Development of Internet Finance', *China Laws Portal*, 2015, https://www-chinajusticeobserver-com.translate.goog/law/x/promoting-sound-development-of-internet-finance-20150718?_x_tr_sl=zh-CN&_x_tr_tl=en&_x_tr_hl=en&_x_tr_pto=sc. Accessed on 25 May 2026.

70 Feng, Emily, 'China's P2P Lenders Brace for More Regulatory Crackdown', *Financial Times*, 2 April 2018, https://www.ft.com/content/d3b00d82-3315-11e8-b5bf-23cb17fd1498. Accessed on 25 May 2026.

71 Bloomberg News, 'China is Set to Start Fresh Round of Checks on P2P Lenders', *Bloomberg*, 20 August 2018, https://www.bloomberg.com/news/articles/2018-08-20/china-is-said-to-start-another-round-of-checks-on-online-lenders. Accessed on 25 May 2026.

72 Yeunai, Xu, 'China Shut Down All P2P Platforms by Mid-Nov', *Asia Times*, 30 November 2020, https://asiatimes.com/2020/11/china-shut-down-all-p2p-platforms-by-mid-nov/. Accessed on 25 May 2026.

73 Advance Intelligence Group, *Atome Financial*, https://www.advancegroup.com/business-units/atome-financial/atome. Accessed on 25 May 2026.

74 This is referred to as white labelling. The product belongs to a different party, which then allows others to put their labels on the product for a small fee. In this case, however, the product is not merely the app itself, but the entire ecosystem around it, including the backend infrastructure and data systems.

It is also important to note that, within the app ecosystem, an app is technically defined through a domain-name-based namespace. For instance, if the RBI were to release an app, it would conventionally fall under a namespace such as *in.org.rbi*—the regulator's domain name in reverse. However, no equivalent restriction exists between the technical namespace and the commercial name displayed to users within the app interface. This makes it possible to create a seemingly new app simply by altering the namespace through straightforward string replacement using automated packaging tools. White labelling through technical means is therefore vastly more scalable than the repackaging of ordinary physical goods.

75 'ED Raids Chinese Loan App Firms and Payment Gateway Firms in Bengaluru, Freezes Accounts Worth ₹17 crore', *The Hindu*, 3 September 2022, https://www.thehindu.com/news/national/karnataka/ed-raids-chinese-mobile-loan-app-firms-and-payment-gateway-firms-in-bengaluru/article65844803.ece. Accessed on 25 May 2026.

76 'ED Conducts Fresh Raids in Chinese Loan Apps Case', *The Indian Express*, 14 September 2022, https://indianexpress.com/article/india/ed-conducts-fresh-raids-in-chinese-loan-apps-case-8151442/. Accessed on 25 May 2026.

77 Directorate of Enforcement, 'Chinese Loan App Case', Press Release, 21 October 2022, https://web.archive.org/web/20221021165042/https://enforcementdirectorate.gov.in/sites/default/files/latestnews/Chinese%2520Loan%2520App%2520Case.pdf. Accessed on 25 May 2026.

78 Gupta, Manoj, 'Chinese Game Plan Uncovered with Loan App Scam, Rs 141 Crore Laundered Via Crypto', *News18*, 1 April 2025, https://www.news18.com/india/chinese-game-plan-uncovered-with-loan-app-scam-rs-141-crore-laundered-via-crypto-exclusive-9282806.html. Accessed on 25 May 2026.

79 Reserve Bank of India, 'RBI Cancels Certificate of Registration of X10 Financial Services Limited due to Irregular Lending Practices', 21 January 2025, https://www.rbi.org.in/scripts/FS_PressRelease.aspx?prid=59565. Accessed on 25 May 2026.

80 Reserve Bank of India, 'RBI Cancels CoR of Zavron Finance', 2 December 2024, https://www.rbi.org.in/scripts/bs_pressreleasedisplay.aspx/BS_PressReleaseDisplay.aspx?prid=59225. Accessed on 25 May 2026.

81 Reserve Bank of India, 'RBI Cancels CoR of Five NBFCs', 25 May 2022, https://www.rbi.org.in/scripts/BS_PressReleaseDisplay.aspx?prid=53763. Accessed on 25 May 2026.

82 'Nagpur: Why Petrol Pump Owners Plan to Halt Digital Payments from May 10', *Nagpur Today*, 4 May 2025, https://www.nagpurtoday.in/nagpur-why-petrol-pump-owners-plan-to-halt-digital-payments-from-may-10/05040752. Accessed on 25 May 2026.

83 Reserve Bank of India, 'RBI Payment System Indicators', August 2025, https://rbidocs.rbi.org.in/rdocs/PSI/PDFs/PSIAUGUST20253E4CD7C42FAE-4896B92ABF9DD425B33D.PDF. Accessed on 25 May 2026.

84 Press Information Bureau, Government of India, 'Under the Leadership of Prime Minister Shri Narendra Modi and Guidance of Home Minister Shri Amit Shah, Building a Cyber Safe India is One of the Top Priorities of the Ministry of Home Affairs', 3 January 2024, https://www.pib.gov.in/PressReleseDetailm.aspx?PRID=1992949. Accessed on 25 May 2026.

85 Reserve Bank of India, 'RBI Payment System Indicators', December 2023, https://rbidocs.rbi.org.in/rdocs/PSI/PDFs/PSIDECEMBER2023DFAE63EFAFB94F759A87DC8ED8DF0211.PDF. Accessed on 25 May 2026. The values from January 2023 to December 2023 in Table 5 are added together to come up with the final figure.

86 LocalCircles, '1 in 5 Families Surveyed that Have Someone Using UPI Confirm Experiencing UPI Fraud Once or More in the Last 3 Years; 51% of Them Did Not File Any Complaint, Anywhere', survey conducted between March and June 2025, https://www.localcircles.com/a/press/page/upi-fraud-complaint. Accessed on 25 May 2026.

87 Infosys Limited, 'Annual Report 2023–24 Analysis', 3 June 2024, https://reports.muthu.co/posts/infosys-ltdannual-report-2023-24-analysis. Accessed on 21 May 2026.

88 TCS, *Integrated Annual Report 2023–24*, p. 128, https://www.tcs.com/content/dam/tcs/investor-relations/financial-statements/2023-24/ar/annual-report-2023-2024.pdf. Accessed on 21 May 2026.

89 Wipro, *Integrated Annual Report 2023–24*, p. 41, https://www.wipro.com/content/dam/nexus/en/investor/annual-reports/2023-2024/integrated-annual-report-2023-24.pdf. Accessed on 21 May 2026.

90 Heffernan, Patrick, 'New IT Services Vertical Revenue Data Shows TCS' Public Sector Surge and Market Shifts', *TBR Insight Center*, 19 February 2025, https://tbri.com/blog/new-it-services-vertical-revenue-data-shows-tcs-public-sector-surge-and-market-shifts/. Accessed on 21 May 2026.

91 Sebastian, Merryl, 'NR Narayana Murthy: Why Indians Are Debating a 70-Hour Work Week', *BBC News*, 1 November 2023, https://www.bbc.com/news/world-asia-india-67269976. Accessed on 21 May 2026.

92 Leaderboard Overview, *Arena*, https://lmarena.ai/leaderboard/. Accessed on 21 May 2026.

93 'EleutherAI /pythia', *Github*, https://github.com/EleutherAI/pythia. Accessed on 21 May 2026.

94 NASSCOM, *The Developer's Playbook for Responsible AI in India*, November 2024, https://nasscom.in/ai/pdf/the-developer's-playbook-for-responsible-ai-in-india.pdf. Accessed on 21 May 2026.

95 Snyder, Alison, and Maria Curi, 'Scoop: Advanced AI Chips Cleared for Export to UAE under Microsoft Deal', *Axios*, 7 December 2024, https://www.axios.com/2024/12/07/us-uae-microsoft-g42-ai-chips. Accessed on 21 May 2026.

96 Quell, Molly, 'Trump's Sanctions on ICC Prosecutor Have Halted Tribunal's Work', *Associated Press*, 15 May 2025, https://apnews.com/article/icc-trump-sanctions-karim-khan-court-a4b4c02751ab84c09718b1b95cbd5db3. Accessed on 21 May 2026.

97 Ropek, Lucas, 'Palantir's Billionaire CEO Just Can't Stop Talking About Killing People', *Gizmodo*, 7 February 2025, https://gizmodo.com/palantirs-billionaire-ceo-just-cant-stop-talking-about-killing-people-2000560597. Accessed on 21 May 2026.

98 Pandey, Mohit, 'IndiaAI's Equity in AI Startups Is a "Fantastic" but Risky Idea', *Analytics India Magazine*, 27 June 2025, https://analyticsindiamag.com/ai-features/indiaais-equity-in-ai-startups-is-a-fantastic-but-risky-idea/. Accessed on 21 May 2026.

99 Ibid.

100 Rao, Vidhatri, 'One Dialect, One Fable at a Time: A Telugu Model Opens a Window for India AI Dream', *The Indian Express*, 14 March 2025, https://web.archive.org/web/20250313020146/https://indianexpress.com/article/india/one-dialect-one-fable-at-a-time-a-telugu-model-opens-a-window-for-india-ai-dream-9883733/. Accessed on 21 May 2026.

101 'Viswam.AI Launches Summer of AI to Train One Lakh AI Engineers and Create World's First Telugu LLM', *Telangana Today*, 13 May 2025, https://telanganatoday.com/viswam-ai-launches-summer-of-ai-to-train-one-lakh-ai-engineers-and-create-worlds-first-telugu-llm. Accessed on 21 May 2026.

102 Nutan C., post on X (formerly Twitter), 20 July 2025, https://x.com/nutanc/status/1946740589162328143. Accessed on 8 June 2026.

103 Perrin, Trevor, and Moxie Marlinspike, *The Double Ratchet Algorithm*, rev. 4, 4 November 2025, https://signal.org/docs/specifications/doubleratchet/. Accessed on 25 May 2026.

104 Barik, Soumyarendra, 'Should the Police Know Who Sent a Personal WhatsApp Message?', *Entrackr*, 23 June 2021, https://entrackr.com/2021/06/should-the-police-know-who-sent-a-personal-whatsapp-message/. Accessed on 25 May 2026.

105 Medeiros, Ben, and Pawan Singh, 'Addressing Misinformation on WhatsApp in India through Intermediary Liability Policy, Platform Design Modification, and Media Literacy', *Journal of Information Policy*, vol. 10, 2020, pp. 276–298, https://www.jstor.org/stable/10.5325/jinfopoli.10.2020.0276. Accessed on 25 May 2026.

106 Gupta, Himanshu, and Harsh Taneja, 'WhatsApp Has a Fake News Problem—That Can Be Fixed without Breaking Encryption', *Columbia Journalism Review*, 23 August 2018, https://www.cjr.org/tow_center/whatsapp-doesnt-have-to-break-encryption-to-beat-fake-news.php. Accessed on 25 May 2026.

107 Green, Matthew, 'Thinking about "Traceability"', *A Few Thoughts on Cryptographic Engineering*, 1 August 2021, https://blog.cryptographyengineering.com/2021/08/01/thinking-about-traceability/. Accessed on 21 May 2026.

108 'Centre Bans 14 Mobile Messenger Apps Allegedly Being Used by Terror Groups', *The Hindu BusinessLine*, 1 May 2023, https://www.thehindubusinessline.com/news/centre-bans-14-mobile-messenger-apps-allegedly-being-used-by-terror-groups/article66800689.ece. Accessed on 25 May 2026.

109 Tiwari, Shubham, 'Apps Banned over Terror Link in 2023 Still Thriving, Government Acknowledges', *India Today*, 28 February 2025, https://www.indiatoday.in/india/story/chinese-banned-apps-returns-terror-link-security-threat-2687139-2025-02-28. Accessed on 25 May 2026.

110 'A Non-Anthropomorphized View of LLMs', 6 July 2025, https://addxorrol.blogspot.com/2025/07/a-non-anthropomorphized-view-of-llms.html

111 Reserve Bank of India, *FREE-AI Committee Report: Framework for Responsible and Ethical Enablement of Artificial Intelligence*, p. 40, para. 4.4.8, https://rbidocs.rbi.org.in/rdocs/PublicationReport/Pdfs/FREEAIR130820250A24FF2D4578453F-824C72ED9F5D5851.PDF. Accessed on 25 May 2026.

112 Vaishnav, Anurag, *Non-Personal Data Governance Framework: Report Summary*, PRS Legislative Research, 28 July 2020, https://prsindia.org/policy/report-summaries/non-personal-data-governance-framework. Accessed on 25 May 2026.

113 Reserve Bank of India, *FREE-AI Committee Report: Framework for Responsible and Ethical Enablement of Artificial Intelligence*, p. 47, para. 4.4.25, https://rbidocs.rbi.org.in/rdocs/PublicationReport/Pdfs/FREEAIR130820250A24FF2D4578453F824C72ED9F5D5851.PDF. Accessed on 25 May 2026.

114 Chan, Kyle, 'China's Overlapping Tech-Industrial Ecosystems', *High Capacity*, 22 January 2025, https://www.high-capacity.com/p/chinas-overlapping-tech-industrial. Accessed on 25 May 2026.